AMERICAN FARM BOOK:

OR

COMPEND OF AMERICAN AGRICULTURE;

BEING A

PRACTICAL TREATISE ON SOILS, MANURES, DRAINING
IRRIGATION, GRASSES, GRAIN, ROOTS, FRUITS,
COTTON, TOBACCO, SUGAR CANE, RICE,

AND EVERY

STAPLE PRODUCT OF THE UNITED STATES.

WITH THE

BEST METHODS

OF PLANTING, CULTIVATING, AND PREPARATION FOR MARKET

ILLUSTRATED BY MORE THAN 100 ENGRAVINGS.

By R. L. ALLEN.

AUTHOR OF DISEASES OF "DOMESTIC ANIMALS," AND EDITOR OF
THE AMERICAN AGRICULTURIST"

THE LYONS PRESS

Guilford, Connecticut

An imprint of The Globe Pequot Press

Originally published in 1849 by Orange Judd & Company

First Lyons Press edition, May 2002

The Lyons Press is an imprint of The Globe Pequot Press.

10 9 8 7 6 5 4 3 2

Printed in Canada

ISBN-13: 978-1-58574-354-4
ISBN-10: 1-58574-354-2

Library of Congress Cataloging-in-Publication Data

Allen, R. L. (Richard Lamb), 1803-1869.
     The American farm book, or Compend of American culture: being a practical treatise on soils, manures, draining, irrigation, grasses, grain, roots, fruits, cotton, tobacco, sugar, cane rice, and every staple product in the United States; with the best methods of planting, cultivating, and preparations for market / by R.L. Allen.-1st Lyons Press ed.
           p. cm.
     "Originally published in 1849 by Orange Judd Company"-T.p. verso.
     ISBN-10: 1-58574-354-2 (alk. Paper)
     Agriculture-United States-Handbooks, manuals, etc.
I. Title: Compend of American Culture. II. Title.

S501.2A58 2002
630'.973-dc21   2001050523

PREFACE.

The following work on American Agriculture is intended as one of the first in the series of lessons for the American Farmer. The size precludes its embracing any thing beyond the shortest summary of the principles and practice by which he should be guided, in the honorable career he has selected. As a primary work, it is not desirable it should comprise so much as to alarm the tyro in agriculture with the magnitude of his subject. A concise and popular exposition of the principal topics to which his attention will necessarily be directed, will, it is believed, in connection with his own observation and practice, give him a taste for further research, which will lead him to the fullest attainment in agricultural knowledge that could be expected from his capacity and opportunities.

Much of what is detailed in the present volume, has been tested by the writer's own experience and observation. For the remainder, he is indebted to various oral and written information, derived from the best agriculturists, and especially from the valuable foreign and domestic agricultural periodicals of the present day.

Whenever *original* authority could be known or recollected, it has been credited; but many even of the most recent discoveries, have already passed through such numerous hands, and received so many shades of alteration or improvement, that their authors would hardly recognize their own offspring. It would not be strange, therefore, if they had become incorporated in the mass of agricultural principles, without any indication of their origin. The same or similar discoveries and improvements, are also not unfrequently made without any interchange, by different minds and at remote distances. If any omissions of proper acknowledgment have occurred, the writer will be happy to correct them hereafter.

To the experienced and scientific, this work may appear too commonplace—to the uninstructed, too enlarged or abtruse. It was not intended to reconcile impossibilities. The first must look to elaborate or complete treatises for the fullest information on the various subjects comprehended in this general summary. To the last, it must be answered, that what is here communicated, is important to be known; that modern agriculture, like all other progressive modern sciences and arts, has necessarily introduced new terms for the explanation of new principles and new practices; and the former must be learned before the latter can be comprehended.

INTRODUCTION

AGRICULTURE, in its most extensive sense, may be defined the cultivation of the earth with a reference to the production of vegetables, and the conversion of portions of them into animals, and a variety of forms, which are the best adapted to the wants of mankind. It is appropriately distinguished by numerous subdivisions.

Tillage Husbandry consists in the raising of grain, roots and other products, which require the extensive use of the plow and harrow to prepare the ground for annual sowing and planting.

Grazing is limited to the pasturing and winter feeding of farm stock, and it requires that the land appropriated to this purpose, should be kept in pasturage for summer food, and in meadows to yield the hay necessary for winter's use. In its strictly technical meaning, grazing implies the rearing of farm stock till they have attained sufficient maturity for a profitable market, as far as this maturity can be secured on grass and hay. It, however, properly embraces in its minor divisions, the keeping of cows for the purposes of a dairy, and the support of flocks for the production of wool.

Feeding, in its agricultural signification, consists in stall fattening animals. It is properly connected with tillage husbandry for the production of grain and roots, and by the free use of which, animals can be brought to a higher condition or *ripeness*, and they will thus command a much better price in market, than if fed exclusively on grass and hay.

Breeding, technically defined, is restricted to the production of choice animals for use as future breeders by the judicious selection and crossing of the best specimens of the various distinct breeds of domestic stock.

Horticulture embraces the entire department of gardening, the cultivation of trees, shrubbery, and fruits; and these occupations are again variously subdivided.

By *Planting* (or the occupation of planters), is understood the cultivation of extensive farms or plantations, for the exclusive production of one or more commercial staples, such as cotton, sugar, rice, tobacco, indigo, &c., and their preparation for a distant market. The term is peculiarly sectional, and its use so far as adopted in this country, is limited to the southern portion of it.

All of the foregoing and various other occupations connected with the cultivation of the earth, are comprehended under the general head of Agriculture.

Besides the varied practical knowledge which is indispensable to the proper management of every department of agriculture, its general principles and theoretical relations require a familiarity with the elements of History, Geology, Meteorology, Chemistry, Botany, Entomology, Anatomy, Zoology, Animal and Vegetable Physiology, and Mechanics; and in their ultimate connection, they involve no inconsiderable share of the entire circle of human knowledge and science.

In view of its intricacy, its magnitude, and its importance to the human race, we cannot fail to be struck with the peculiar wisdom of Deity in assigning to man this occupation, when a far-seeing and vigorous intellect fitted him to scan with unerring certainty and precision, the visible works of his Creator, and trace their causes and effects through al. their varied relations. It was while in the sinless perfection of his original nature, when " the Lord God put him into the garden of Eden, to dress it and to keep it," and agriculture was his his sole occupation, that his godlike intelligence enabled him, instinctively to give appropriate names, indicative of their true nature or character, "to all cattle, and to the fowl of the air, and to every beast of the field;" and so just and accurate was his perception, that ' whatsoever he called every living creature, that was the name thereof."

In our present imperfect condition, a beneficent Providence has not reserved a moderate success in Agriculture, exclusively to the exercise of a high degree of intelligence. His laws have been so kindly framed, that the hand even of uninstructed toil, may receive some requital in remunerating harvests; while their utmost fulness can be anticipated, only where corporeal efforts are directed by the highest intelligence.

The indispensable necessity of an advanced agriculture to the comforts and wealth, and indeed, to the very existence of a great nation, renders it an object peculiarly worthy the attention and regard of the legislative power. In looking to the history both of ancient and modern times, we find, that wherever a people have risen to enduring eminence, they have sedulously encouraged and protected this right arm of their strength. Examples need not be given, for they abound in every page of their civil polity.

Our own country has not been wanting in a moderate regard for Agriculture. By wise legislation in our National Congress, every item of extensive agricultural production within the United States, with the single exception of the inferior wools, is believed to be fully protected from foreign competition, by an unyielding and perfectly adequate impost on all such articles, as would otherwise enter into a successful rivalry with them from abroad. Many of our subordinate, or state legislatures, have also, by liberal provisions,

given such encouragement to various objects, as they deemed
necessary to develope the agricultural resources within their
jurisdiction.   Such have been the appropriations for numerous geological and other state surveys; the bounties on dif
ferent articles, as silk, hemp, and some others ; and occasionally a small gratuity to encourage the formation and
support of State and County Agricultural Societies.

But while we would not be unmindful of what has heretofore been effected, our duty compels us to assert, that much
yet remains to be done.   A single suggestion for the action
of the general government and states, is all that our limits
will permit us to make.

The organization of a National Board of Agriculture, composed of able and intelligent men, expressly selected for this
purpose, whose sole duty it should be, to collect all information and statistics on the subject, and arrange and spread
them before the people ; to introduce new and valuable foreign plants, adapted to our soil and climate ; suggest improved methods of cultivation ; recommend and disseminate
the most approved principles of breeding domestic animals ;
indicate those best adapted to particular purposes or peculiar
localities ; point out new avenues for the profitable disposal
of our surplus products ; and recommend such laws or their
modification, as might best subserve this interest ; in short,
who should stand as sentinels and defenders on the watch-
tower of this great citadel—this is the lofty duty, and
should be esteemed the peculiar privilege of American legislation to accomplish.   This was a favorite, yet not a fully
digested plan of Washington, the suggestions of whose be-
nevolent and comprehensive mind were never followed but
for his country's good.

From the individual states, a less commanding, but not
less beneficial duty is required.   Restrictions wisely imposed upon the general government, limit its action to such

measures only as are essential to the general welfare, and such as cannot properly be accomplished by any more circumscribed authority. More liberal and enlarged grants from the people, give to the state legislatures the power of doing all which their constituents choose to have effected for their own benefit.

Education in all its branches, is under their exclusive control ; and to endow and foster every institution which has a tendency to raise and improve the intellectual, the moral, and the social condition of the people, has ever been their cherished policy. Yet up to this time, no institution expressly designed for the professional education of farmers, has ever been established in this country. That far-seeing wisdom which characterizes the consummate statesman, and which regards the future equally with the present and past, has halted upon the threshold of the great temple of agricultural science, whose ample and enduring foundations have been commenced by the united efforts of the men of genius throughout both hemispheres. To aid with every means in their power in laying these foundations broad and deep, to elevate its superstructure, to rear its mighty columns, and adorn its graceful capitals, would seem most properly to come entirely within the province of the representatives of intelligent freemen, the great business of whose lives is the practice of agriculture.

In addition to continuing and making more general and comprehensive the encouragement for other objects heretofore considered, it is the duty of each of the larger states of the Union, liberally to endow and organize an Agricultural College, and insure its successful operation within its jurisdiction. Connected with these, should be example and experimental farms, where the suggestions of science should be amply tested and carried out before submitting them to the public. The most competent men at home and abroad should be invited to fill a professional chair; and if money would tempt a Liebig, a Boussingault, a Johnston, or a Playfair, to leave the investigations of European soils and products, and devote his mind and energies to the improvement of American husbandry, it should be freely given.

These institutions should be schools for the teachers equally with the taught; and their liberally-appointed laboratories and collections should contain every available means for the discovery of what is yet hidden, as well as for the further development of what is already partially known. Minor institutions should of course be established at different and remote points, to scatter the elements of agricultural knowledge broadcast over the land, and bring them within the reach of the poorest citizens and the humblest capacities.

By such a liberal and enlightened course, we should not only incalculably augment the productive agricultural energies of our own country, but we should also in part, repay to the world at large, the obligations under which we now rest for having appropriated numerous and important discoveries and improvements from abroad. If we have the ability which none can doubt, we should make it a point of honor to return in kind, the liberal advances we have thus received.

It is to the rising generation these suggestions are made; the risen are not yet prepared for their acceptance. The latter have been educated, and become habituated to different and more partial influences. By their industry, intelligence, and energy, displayed in numberless ways, and especially by their protection of American labor, they have accomplished much for their own and their country's welfare—they are resolved to leave this glory for their successors.

New York, *June,* 1846

AMERICAN AGRICULTURE.

CHAPTER I.

SOILS.

Soils are those portions of the earth's surface, which contain a mixture of mineral and vegetable or animal substances, in such proportions as adapt them to the support of vegetation. Rocks are the original basis of all soils, which by the convulsions of nature, or the less violent but long continued and equally efficient action of air, moisture and frost, have been broken into fragments more or less minute. There are various gradations of these changes.

THE TEXTURE OF SOILS.—Some rocks exist in large boulders or rounded stones, which thickly overspread the surface and mingle themselves with the earth beneath, giving to it the character of a rocky soil. The smaller sizes, but an equal prevalence of the same materials, distinguish the surface where they abound, as a stony soil. A third and more minute division is called a gravelly soil; a fourth is a sandy soil; a fifth constitutes a loam; and a sixth, in which the particles of earth are reduced to their greatest fineness, is known under the name of a clay soil.

The two first mentioned, however, are not properly, distinct soils, as the only support of any profitable vegetation, is to be found in the finer earth in which the rocks and stones are embedded. In frequent instances, they materially benefit the crops, by the influence produced from their shade, moisture, and protection from winds; and by the gradual decomposition of such as contain lime, potash and other fertilizing materials, they enrich the soil and contribute to the support of vegetation. Their decomposition is hastened by the apparently worthless vegetable life which they yield to the living mosses that cling to their sides, and every where

penetrate their fissures; thus imperceptibly corroding the
solid structures and preparing them for future usefulness as
soils. If we add to the above, a peat or vegetable soil, we
shall have the material divisions of soils, as distinguished by
their texture.

Besides these, soils are frequently to be seen, more or less
extensive, which possess peculiarities entitling them to a
distinct classification, and requiring a treatment, in some
respects, different from any others. Such are the *prarie*
soils, which, having been annually burnt over for ages, are
highly charged with ashes and the alkaline salts. Such
also, are the *terre-gras* lands in Louisiana, and the *brick-
mold* of the East and West Indies, each of which requires
peculiar management in plowing and cultivation.

OTHER CLASSIFICATIONS OF SOILS.—Soils are also distinguished by their tendency to absorb and retain water,
gravel and sand holding very little, while clay and peat
readily absorb and retain a great deal; by their constant saturation from perennial springs, which are called springy soils;
by the quantity of vegetable and animal matter they contain;
by their porosity or adhesiveness; by their chemical character, whether silicious, argillaceous or calcareous; by the
quality and nature of the vegetation they sustain; and lastly,
and by far the most important, they are distinguished by their
fertility or barrenness, the result of the proper adjustment and
combination of most of the conditions enumerated. Deserts
of sands, layers of rocks, stone or pure gravel, and beds of
marl and peat are not soils, though containing many of their
most important elements.

It is apparent to the most casual observer, that soils frequently, and by almost imperceptible degrees, change from
one character to another, and that no classification, however
minute, will suffice to distinguish each. Some obvious yet
simple distinctions, which are usually recognized, must
nevertheless be assumed for future reference. For this purpose, and to avoid unecessary deviations from what should
be a common standard, we shall adopt the arrangements as
made by Professor Johnston, which is based principally
upon their chemical constituents.

"1°. *Pure clay* (pipe-clay) consisting of about 60 of silica
and 40 of alumina and oxide of iron, for the *most part* chemically combined. It allows no silicious sand to subside when
diffused through water, and rarely forms any extent of soil.

2°. *Strongest clay soil* (tile-clay, unctuous clay) consists of pure clay mixed with 5 to 15 per cent of a silicious sand, which can be separated from it by boiling and decantation.

3°. *Clay loam* differs from a clay soil, in allowing from 15 to 30 per cent. of fine sand to be separated from it by washing, as above described. By this admixture of sand, its parts are mechanically separated, and hence its freer and more friable nature.

4°. A *loamy soil* deposits from 30 to 60 per cent. of sand by mechanical washing.

5°. A *sandy loam* leaves from 60 to 90 per cent. of sand, and

6°. A *sandy soil* contains no more than ten per cent. of pure clay.

The mode of examining, with the view of naming soils, as above, is very simple. It is only necessary to spread a weighed quantity of the soil in a thin layer upon writing paper, and to dry it for an hour or two in an oven or upon a hot plate, the heat of which is not sufficient to discolor the paper—the loss of weight gives the water it contained. While this is drying, a second weighed portion may be boiled or otherwise thoroughly incorporated with water, and the whole then poured into a vessel, in which the heavy sandy parts are allowed to subside until the fine clay is beginning to settle also. This point must be carefully watched, the liquid then poured off, the sand collected, dried as before upon paper, and again weighed. This weight is the quantity of sand in the known weight of *moist* soil, which by the previous experiment has been found to contain a certain quantity of water.

Thus, suppose two portions, each 200 grs., are weighed, and the one in the oven loses 50 grs. of water, and the other leaves 60 grs. of sand,—then, the 200 grs. of *moist* are equal to 150 of *dry*, and this 150 of dry soil contain 60 of sand, or 40 in 100 (40 per cent.). It would, therefore, be properly called a *loam*, or *loamy* soil.

But the above classification has reference only to the clay and sand, while we know that lime is an important constituent of soils, of which they are seldom entirely destitute. We have, therefore,

7°. *Marly soils*, in which the proportion of lime is more than five but does not exceed 20 per cent. of the whole weight of the dry soil. The marl is a sandy, loamy, or clay marl, according as the proportion of clay it contains would place it under the one or other denomination, supposing it to be

entirely free from lime, or not to contain more than five per
cent., and

8°. *Calcareous soils*, in which the lime exceeding 20 per
cent. becomes the distinguishing constituent. These are
also calcareous clays, calcareous loams, or calcareous sands
according to the proportion of clay and sand which are
present in them.

The determination of the lime also, when it exceeds five
per cent., is attended with no difficulty.

To 100 grs. of the dry soil diffused through half a pint of
cold water, add half a wine glass-full of muriatic acid (the
spirit of salt of the shops), stir it occasionally during the day,
and let it stand over night to settle. Pour off the clear
liquor in the morning and fill up the vessel with water, to
wash away the excess of acid. When the water is again
clear, pour it off, dry the soil and weigh it—the loss will
amount generally to about one per cent. more than the quantity of lime present. The result will be sufficiently near,
however, for the purposes of classification. If the loss
exceed five grs. from 100 of the dry soil, it may be classed
among the marls, if more than 20 grs. among the calcareous
soils.

Lastly, vegetable matter is sometimes the characteristic
of a soil, which gives rise to a further division of

9°. *Vegetable molds*, which are of various kinds, from
the garden mold, which contains from five to ten per cent.,
to the peaty soil, in which the organic matter may amount
to 60 or 70. These soils also are clayey, loamy, or sandy,
according to the predominant character of the earthy ad-
mixtures.

The method of determining the amount of vegetable
matter for the purposes of classification, is to dry the soil well
in an oven, and weigh it; then to heat it to dull redness over
a lamp or a bright fire till the combustible matter is burned
away. The loss on again weighing is the quantity of organic
matter."

The foregoing are only such general divisions, as possess
properties sufficiently common to each, to require a treatment
nearly similar. Besides their principal component parts,
every soil must contain in greater or less quantities, all the
elements which enter into the composition of vegetables.
They may have certain substances which are not necessary
to vegetable life, and some one or all of such as are, may be
contained in excess; yet to sustain a healthy prolific vegeta

tion, they must hold, and in a form fitted to its support, *silex, alumina, carbonate of lime, sulphate of lime, potash, soda, magnesia, sulphur, phosphorus, oxide of iron, manganese, chlorine,* and probably *iodine.* These are called the inorganic or earthy parts of soils, as they are found almost exclusively in combination with earths, salts, or minerals. They however, constitute from less than 0.5 (one half of one) to over 10 per cent. of all vegetables. In addition to these, fertile soils must also contain *carbon, oxygen, nitrogen* and *hydrogen,* which are called the organic parts of soils, from their great preponderance in vegetables and animals, of which they constitute from about 90, to over 99 per cent. of their entire substance.

CLAY SOILS—THEIR CHARACTERISTICS AND TREATMENT —Clay soils are usually denominated cold and wet, from their strong affinity to water, which they generally hold in too great excess for rapid or luxuriant vegetation. The alumina which exists in clay, not only combines with water forming a chemical compound, but the minute division of its particles and their consequent compactness, oppose serious obstacles to the escape of such as comes in contact with it. Hence, the necessity of placing it in a condition to obviate these essential defects.

The most effectual method of disposing of the surplus water in clay soils, is by underdraining. This draws off rapidly, yet by imperceptible degrees, all the excess of water, and opens it to the free admission of atmospheric air; and this, in its passage through the soil, imparts heat and such of the gases it contains, as are useful in sustaining vegetation. When these are not constructed, open drains should be formed wherever water stands after rains. The slight elevation and depression of the surface made by careful plowing, will probably be sufficient, if they terminate in some ravine or artificial ditch, and have size and declivity enough to pass off the water rapidly.

Clay soils are greatly improved by coarse vegetable manures, straw, corn-stalks, chips, &c., which tend to the separation of its particles. The addition of sand is very beneficial, but this is too expensive for large fields. Lime is also a valuable material for a clay soil, as by the chemical combinations which are thereby induced, the extreme tenacity of the soil is broken up; while the lime adds an ingredient of fertility, not before possessed by it, perhaps, to an adequate extent. Gypsum has the same effect, in a more powerful degree. Paring and burning, (by which, the surface con taining vegetable matter, is collected into heaps and fired, reducing the mass to a charred heap, which is then spread over and mixed with the soil,) produce the same result. This is a practice which has been long in use in different parts of Europe; but although attended with immediate and powerful results, it is too expensive for general introduction into a country, where labor is high, and land and its products comparatively cheap.

Wherever frosts and snow abound, the plowing of clay lands for spring crops, should be done in the autumn if practicable; by which their adhesiveness is temporarily destroyed, the earth is enriched by the snows, and finely pulverized by the frost, and they are left in the finest condition for early spring sowing, and without additional working. If plowed in the spring, it should be done when they are neither too wet nor dry; if the former, the earth subsequently bakes, and for a long time, it is almost impenetrable to the hoe or the teeth of the harrow; if too dry, they are so compact as to be turned over only with great effort, and then in solid lumps. The action of the atmosphere, will pulverize these masses of baked earth after a time; but not sufficiently early in most of our northern states, for the convenience or advantage of such crops as are immediately to follow the plowing. For much of the South, plowing clay lands in the autumn is worse than useless; as the loose earth thus thrown up, is soon reduced by the heavy winter rains to a compact surface, apparently as unfitted for cultivation, without subsequent plowing, as the incrustations of lava from a volcano.

No soils are so tenacious of the manures which may be incorporated with them as the clays. They form an intimate combination, both mechanical and chemical,* and hold

---

* By *mechanical*, in the sense above used, is understood the external relation of bodies, which is nearly equivalent in its meaning in this connection, to artificial. Thus the clay envelopes the manure, and from its impervious character shields it from escape either by drainage or evaporation, and almost as effectually, as if it were enclosed in an earthen vessel.

By *chemical* is meant, its internal or constitutional character. Thus clay not only absorbs the gases which are brought into contact with it from manures, from moisture and from air, as a sponge absorbs water; but it also forms new combinations with them, which change the original nature of these elementary principles, and from light evanescent gases, they become component parts of solid bodies, in which condition they are retained till exhausted by the growing vegetation.

These terms are important, and should be fully understood. For

them securely against waste from drainage or evaporation
tor an indefinite time, till the growing crops demand them.
They also greedily seize upon and hoard up all such fertilizing principles as are conveyed to them by the air and
rains. We may mention as an example of their efficiency
in abstracting vegetable nutrition from the atmosphere, that
many of them, when thrown out from a great depth below
the surface, and entirely destitute of organic remains (vegetable or animal matter), after an exposure for some months
to its meliorating influence, become capable of bearing large
crops, without the aid of manure. This is particularly true
of the clays which rest on the Onondaga limestone, an extensive group occupying the central and north-western part
of New York.

The clays are admirably adapted to the production of
most of the grains, and the red and white clovers cultivated
in the United States. These they yield in great profusion
and of the best quality; and so peculiarly suited are they to
permanent meadows and pasturage, that they are styled by
way of eminence, *grass lands.* They are justly character-
ised as strong and lasting soils; and when properly managed

---

the purpose of still more clearly elucidating the subject to the mind
of the young student, we give some further examples. If we take a
piece of crystalized marble, compact uncrystalized limestone, and
chalk, we shall have three substances exactly alike in their *chemical*
character; for they are all chemical combinations of carbonic acid and
lime, associated together in precisely the same proportions. But in
their external arrangements, as they appear in a recent fracture to the
eye and touch, that is, in their *mechanical* arrangements, they are
totally dissimilar.

Again—If we take the pure lime, (quick lime), that is obtained from
each of the foregoing by subjecting them to an intense heat, by which
the carbonic acid is expelled, and pour upon it nearly one third of its
weight of water, great heat is developed, and the lime both *mechanically* absorbs, and *chemically* combines with it, forming a new compound,
or salt, which is a hydrate of lime.

If sand (mostly silex) be added to the lime with water, and *mechanically* mixed or stirred together and allowed to remain for a sufficient
time, they will combine *chemically,* fo·ming silicate of lime, the
common mortar of stone masons.

Sand (silex) stirred in with clay, (an impure alumina), is *mechanically* mixed; if then subject to a strong heat as in making brick, they
become *chemically* united, forming silicate of alumina, inseparable by
any human means short of the chemists crucible. If we divide or
separate a stick by splitting or cutting, it is a *mechanical;* and if by
burning or charring, it is a *chemical* change. Thus every alteration,
either in nature or art, is referable to one of the above conditions or
changes

and put to their appropriate use, they are esteemed as among the choicest of the farmer's acres.

SANDY SOILS AND THEIR MANAGEMENT.—The character and treatment of sandy soils, are in almost every particular the reverse of those of clay. They do not possess the property of adhesiveness, and they have but little affinity for water, which escapes from them almost as soon as it falls. They have but a slight hold upon the manures which are diffused through them ; they are loose in their texture, and may be plowed at any time, but with most advantage when wet. The sowing or planting should follow immediately.

As clay soils are much benefited by a mixture of sand, so likewise are sandy soils greatly improved by the addition of clay, yet in a much higher degree ; for though it would never pay, as a general rule, to add sand to clay, yet the addition of a few loads of the stiffest clay to a light sand, would in almost every instance, much more than compensate for the trouble and expense. For this purpose, the clay should be thinly spread in autumn, upon sward land previously plowed, and the winter's frost will effectually separate the particles. It should then be harrowed thoroughly and deeply in the spring, and subsequently plowed if necessary. Such a dressing on a light crawling sand, is more than equivalent to an equal quantity of the best manure, and will be permanent in its effects. Clay and sand are necessary to each other, as they both contain qualities which are essential to a good soil ; and that will always be found the best, which has the proper proportion of each.

Sandy soils are improved by the frequent use of a heavy roller ; it cannot be used too often. They require to be made more compact, and any treatment that secures this object, will be advantageous.

Lime, by its chemical action on the constituents of soils, while it separates clay, renders sand more adhesive ; and when cheaply obtained, it is always a profitable dressing for sandy soils, to the full amount they may require. Gypsum, in considerable quantities, has an effect similar to lime, both on clay and sand ; and when added in smaller portions, produces a striking increase in the crops of sandy soils. Clay marls, containing either carbonate, sulphate, or phosphate of lime, are of great value to sandy soils. Equally beneficial are ashes, leached or unleached, peat, or vegetable manures of any kind. Some calcareous sands, containing a large proportion of lime, like those of Egypt and extensive regions in the Barbary States, will produce luxuriantly, if supplied with a slight addition of manure and an abundance of water. Sandy soils can never be profitably cultivated, till they have acquired sufficient compactness and fertility, to sustain a good growth of grass or clover; and when once brought to this condition, they are among the most valuable for tillage, especially for such crops as require early maturity.

They are, at all times, easily plowed and worked; they require no draining; and though light and dry, are quick and kindly soils, giving an immediate and full return for the labor and manure bestowed upon them. When in a condition to produce grass, sheep are admirably adapted to preserve and augment their fertility, and by their incessant migrations over it, their sharp hoofs pack the surface closely, producing the same effect as the roller.

GRAVELLY SOILS are in some respects similar to sand, but much less desirable, being appropriately termed *hungry*. Like the latter, they are peculiarly leachy, but in an increased degree, permitting the rapid escape of manures, both by evaporation and drainage. Such as are calcareous or composed of limestone pebbles, are in a great measure not subject to these objections; as the disposing affinities of the lime, (of which enough will be found to exist in the soil in a finely comminuted or divided state, and in this condition is enabled to act efficiently,) have a tendency to retain the vegetable matters, thus compacting the soil, and holding whatever food of plants may from time to time be given to it, for the wants of future crops. Unless of this latter description, gravelly soils should not be subjected to tillage, but appropriated to pasturage, when sheep will keep them in the best and most profitable condition of which they are capable.

LOAMY SOILS being intermediate between clay and sand, possess characteristics, and require a treatment approximating to one or the other, according to the predominance of either quality. They are among the most desirable soils for the various purposes of agriculture.

MARLY AND CALCAREOUS SOILS have always a full supply of lime, and like the loams, they frequently incline towards a clay or sand, requiring a management corresponding to their character. Putrescent and vegetable manures increase their fertility, and these are held with great tenacity till exhausted by crops. In durability or lastingness, they cannot be exceeded, and few are more profitable for cultivation or grazing.

ALLUVIAL SOILS, are such as have been formed from the washing of streams. They vary in their characteristics, from a mixed clay to an almost pure sand; but generally, they combine the components of soils in such proportions as are designated by *loamy soils*, or *sandy loams*. When thus formed they are exceedingly fertile; and if subject to the an-nual overflow of a stream, having its sources far above them, they usually receive such an addition to their productiveness, as enables them to yield large crops perpetually, without further manuring.

They are for the most part easily worked, and are suited to the various purposes of tillage and meadows; but when exposed to overflowing, it is safer to keep them in grass, as this crop is less liable to injury by a freshet; and where sub-ject to washing from the same cause, a well-matted sod is the best protection which can be offered against it. Many of the natural grasses which are found in these meadows, yield a fodder of the highest value.

PEATY SOILS. These are composed almost wholly of peat, and are frequently called vegetable soils. They are exten-sively diffused between the latitudes of 40° and 60° north, at a level with the ocean, and are frequently found in much lower latitudes, when the elevation of the surface produces a corresponding temperature. They generally occupy low swampy levels, but sometimes exist on slight, northern declivities, where the water in its descent is arrested by a succession of basin-shaped cavities.

Their peaty character is acquired, by the growth and par-tial decay, through successive ages, of various aquatic plants, the principal being the sphagnums and lichens. In swamps, many of which were probably small lakes in their origin, the peat is found of an unknown depth, reaching in some instances, beyond 30 and 40 feet. On declivities and occa-sional levels, the peat is sometimes only a few inches in thickness. It is of a blackish or dark brown color, and exists in various stages of decay, from the almost perfect state of fallen stumps and leaves, to an imperfectly defined, ligneous mass, or even an impalpable powder.

In its natural state, it is totally unfit for any profitable vegetation, being saturated with water, of an antiseptic na-ture, which, for an almost indefinite time, resists putrefaction or decay. When thrown out of its native bed and exposed to drain for a few months, much of it is fit for fuel; and it is always of advantage to the muck heaps, as an absorbent of the liquid and gaseous portions of animal and other volatile manures; or it is of great utility when applied alone to a dry, gravelly or sandy soil

Cultivation of Peat Soils. When it is desirable to cultivate a peaty soil, the first process is to drain it of all the moisture which has given to it, and sustained its present character. The drains must be made sufficiently near to each other, and on every side of the bed; or they must, at least, be so located as effectually to intercept and carry off all the springs or running water which saturates the soil; and they should be deep enough to prevent any injurious capillary attraction of the water to the surface. When it has been thoroughly drained, the hommocs if any, must be cut up with the mattoc or spade, and thrown into heaps, and after they are sufficiently dried, they may be burned, and the ashes scattered over the surface. These afford the best top dressing it can receive. Sand or fine gravel, with a large quantity of barn-yard manure and effete lime, should then be added. On some of these, according as their composition approaches to ordinary soils, good crops of oats, corn, roots, &c may be grown; but they are better suited to meadows, and when thus prepared, they will yield great burthens of clover, timothy, red top, and such of the other grasses as are adapted to moist soils. Subsequent dressings of sand, lime, manure and wood ashes, or of all combined, may be afterwards required, when the crops are deficient, or the grasses degenerate.

Peat contains a large proportion of carbon, and the silicates in which such soils are deficient, (and which they procure only in small proportions from the farm-yard manure, but more largely from the sand or gravel,) are essential to be added, in order to furnish an adequate coating for corn stalks, straw and the valuable grasses. As they are exhausted, they must be again supplied or the crops will fail. Besides yielding an important food to the crop, lime is essential to produce decomposition in the mass of vegetable matter, as well as to combine with and aid in furnishing to the growing plants, such of their food as the atmosphere contains. Ashes are among the best applications, as they possess the silicates, lime, potash, and other inorganic materials of plants in great abundance, and in a form readily adapted to vegeta-

ble nutrition.   Gypsum is also a valuable manure for peaty
soils.

SUBSOILS AND THEIR MANAGEMENT.

The efficiency of soils in producing good crops, depends
much on the subsoil. If this consists of impervious clay or
hard-pan, which prevents the drainage of the water, it is
evident, the accumulation of heavy rains will materially injure the vegetation above; for it is certain, that while nothing is more essential to productiveness than an adequate
supply of moisture for the roots, nothing is more injurious
than their immersion in stagnant water.   If this description
of subsoil be deep, the only remedy is thorough underdraining ; if shallow, the crust may be broken up with the subsoil
plow and gradually mixed with the surface soil, when the
water will readily escape below.

A variety of plows have been constructed for this purpose ;
but unless it be intended to deepen the soil by an admixture
of manures, they must not be used for bringing up the
subsoil too rapidly, to mix with that on the surface.   In addition to the more ready escape of water, thus secured by
breaking it up, the air is also admitted, which enables the
roots to strike deeper, and draw their nourishment from a
much greater depth.   The increased distance through
which the roots penetrate, furnishes them with additional
moisture during a season of drought, thereby securing a luxuriant crop when it might otherwise be destroyed.   This is
frequently a great item in the profit of the farmer ; as, besides the increase of crop which follows a dry, hot season,
when a full supply of moisture is furnished, the product is
usually of better quality; and the general deficiency of agricultural produce, which ensues from seasons of drought,
makes his own more valuable.

As a result of this practice, there is also a gradual increase
in the depth of the soil; as the fine and more soluble particles of the richer materials above, are constantly working
down and enriching the loosened earth below.   In time,
this becomes good soil; and this, in proportion to its depth,
increases the area from which the roots derive their nutriment.   So manifest are the advantages which have followed
the use of subsoil plows, that they have been extensively
introduced of late years, among the indispensable tools of
the better class of agriculturists.

When the subsoil is loose and leachy, (consisting of an excess of sand or gravel,) thereby allowing the too ready escape of moisture and the soluble portions of manures, the subsoil plow is not only unnecessary, but positively injurious. In this case, the surface soil should be somewhat deepened by the addition of vegetable matters, so as to afford a greater depth through which the soluble manures must settle, before they can get beyond the reach of the roots; and the supply of moisture would thereby be much augmented. It is better, however, to keep lands of this character in wood or permanent pasture. They are at best, ungrateful soils, and make a poor return for the labor and manure bestowed upon them.

If there be a diversity in the character of the surface and subsoils, one being inclined to sand and gravel, and the other to marl or clay, a great improvement will be secured, by allowing the plow to reach so far down as to bring up and incorporate with the soil, some of the ingredients in which it is wanting. This admixture is also of remarkable benefit in old or long-cultivated fields, which have become deficient in inorganic matters, and in their texture.

The effect of long continued cultivation, besides exhausting what is essential to the earthy part of plants, is to break down the coarser particles of the soil, by the mechanical action of the plow, harrow, &c.; and in a much more rapid degree, by the chemical combinations, which cultivation and manuring produce. A few years suffice to exhibit striking examples in the formation and decomposition of rocks and stones. Stalactites and various specimens of limestone, indurated clays, sandstone and breccias or pudding stones, are formed, in favorable circumstances, almost under our eye; while some limestones, shales, sandstones, &c., break down in large masses annually, from the combined effect of moisture, heat, and frost. The same changes, on a smaller scale, are constantly going forward in the soil, and much more rapidly while under cultivation. The general tendency of these surface changes, is towards pulverisation The particles forming the soil, from the impalpable mite of dust to the large pebbles, and even the stones and rocks are continually broken up by the combined action of the vital roots, and the manures incorporated with the soil, by which new elements of vegetable food are developed and become available, and in a form so minute as to be imbibed by the spongioles of the roots; and by the absorbent vessels, they

2

are afterwards distributed in their appropriate places in the plant. Where this action has been going on for a long period, a manifestly beneficial effect has immediately followed, from bringing up and mixing with the superficial earth, portions of the subsoil which have never before been subject to cultivation.

A subsoil which is permeable by water, is sometimes imperceptibly beneficial to vegetation, not only by allowing the latent moisture to ascend and yield a necessary supply to the plants; but a moisture frequently charged with lime and various other salts, which the capillary attraction brings from remote depths below the surface. It is probably from this cause, that some soils produce crops far beyond the yield which might be reasonably looked for, from the fertilizing materials actually contained in them. This operation is rapidly going forward during the heat of summer. The water thus charged with saline matters, ascends and evaporates at and below the surface, leaving them diffused throughout the soil. After long continued dry weather, a thin, whitish coating of these salts, is frequently discernible on the ground. The enriching effect of these deposites, is one of the compensating results, seldom discovered or acknowledged perhaps, yet wisely designed by a beneficent Providence, to secure a future and increased fertility from the temporary loss occasioned by drought.

Where rain seldom or never falls, this result is noticeable in numerous, and sometimes extensive beds of quiescent (not shifting) sand. Deposits ofttimes occur several inches in thickness. Such are the extensive beds of impure muriate of soda and other salts, in the arid deserts of California; in the southern parts of Oregon; the nitrates found in India, Egypt, Peru, and various other parts of the world.

ADDITIONAL PROPERTIES OF SOILS.

Besides the qualities of soils already noticed, there are several physical conditions which affect their value. They should be of sufficient depth, friable or easily pulverized; they should possess the right color, and be susceptible of the proper admission and escape of heat, air and moisture.

Jethro Tull, who wrote more than a century ago on the subject of agriculture, maintained that if a soil be worked to a proper depth, and perfectly well pulverized, nothing more is necessary to insure an indefinite succession of the most uxuriant crops, without the aid of manures; and it must be confessed, his practice gave some apparently strong confirma tions of his theory. By carrying tillage far below the surface, thus securing the minute division of the earth to a great depth, and rendering it permeable to the roots, he insured the free access of air and moisture, which are among the first and most important requisites in the growth of vege- tables.

But Tull wrote before agriculture became a science, and omitted to estimate the large amount of fertile ingredients, which every crop takes out of the soil, and which can only be supplied by the addition of fresh materials. A succession of crops would therefore, so far reduce the soil, as to render it necessary to add manures, or vegetation must inevitably fail. This careful, laborious practice, could only, for the time being, enhance the crop and prolong its available supplies; yet in accomplishing even this object, his example is worthy of imitation by every tiller of the soil.

Friableness of the Soil is a quality equally removed from the adhesiveness of strong clay, or the openness of loose sand. Good loam, and fertile, alluvial soils always possess this property. When stirred by the plow, the spade or the hoe, the earth ought to fall and crumble readily, although it should be wet. Such a condition secures a ready admis- sion to the roots, which thus easily pervade the soil, and draw their necessary support from it in every direction Under draining, and the addition of coarse manures to clay, fermented manures and ashes to sand, and lime and gypsum to both, will materially enhance their friableness.

Color is an essential feature in soils, and like friableness, it has an important relation to their capacity for heat and moisture. Dark-colored earths, and black in the highest degree, absorb heat more rapidly than any other when exposed to a temperature above their own; and it escapes with equal readiness when their relative temperature is reversed.

A rough pulverised surface, which is seen in the minute inequalities of a friable and well cultivated soil, produces the same result. During the heat of the day, and especially when the sun's rays fall upon the earth, the dark, friable soil imbibes the heat freely, and transmits it to the remotest roots; thus securing that warmth to the plant, which is one of the necessary conditions of its growth. When the temperature of the air falls on the approach of evening, a reversed action in the soil takes place, by which the heat as rapidly escapes

This immediately brings the surface to *the dew point* and secures a copious deposit of moisture, which a friable soil speedily conveys to every part of the roots.

The dew point is attained when the surface of any object is below the temperature of the surrounding air. The careful observer will not fail to discover the formation of dew, not only for some time after the sun has risen, and long before he sinks below the horizon, when the condition above indicated exists; but sometimes even in the fervor of a mid-day sun, when the thick corn, or any luxuriant vegetable growth repels his fierce rays from the earth. In many instances, the rank, dark growing crops themselves, when shielded from the sun's rays by their overspreading tops, become rapid condensers of atmospheric vapor, and the plant drinks in the wholesome and nutritious aliment at every pore, and frequently collects a surplus, which streams down its sides to the thirsty soil beneath. The principle is further illustrated, by the deposit of moisture in large globules on the surface of any object in the shade, which is sensibly below the surrounding temperature ; as is shown by an earthen or metallic vessel filled with cold water, and set in a warm room on a summer's day.

The proper capacity of soils for imbibing and parting with moisture, gives them another decided advantage over others, which have it in an imperfect degree ; as it is found by recent experiments, that rich, porous soils, which are readily penetrated by water and air, absorb the nutritious gases, (oxygen, nitrogen, and their compounds, nitric and carbonic acid, ammonia, &c.) largely from the atmosphere ; and that they do this to an appreciable extent, *only while moist.* The effect of this will readily be estimated, from the well-known, beneficial influence exerted on the growing plant, by the presence of these important elements.

Light colored clays, marls, and sands, are neither in their mechanical texture, friableness, or color, the best suited to promote the growth of plants. *Peat soils,* from a deficiency of inorganic materials, and their too great affinity for water in their natural condition, are even less adapted to the object than either of the preceding.

Schubler has found, that during 12 hours in the night when the air was moist, 1,000 lbs. of entirely dry quartz,*

---

* Quartz, as analyzed by Bergman, gave 93 per cent. of silex ; 6 of alumina ; and 1 of oxide of iron. It comes so near a pure silica, that in treating of it agriculturally, we speak of it as silex or silica.

oi common sond did not gain a pound; calcareous sand gained 2 lbs.; loamy soil, 21 lbs.; clay loam, 25 lbs.; such as were rich in vegetable mold, still more; while peat absorbed a much larger per cent. than either.

Davy also found, that the same quantity of very fertile and perfectly dry soil, on exposure, gained 18 lbs. in one hour; a good sandy soil, under the same circumstances, absorbed 11 pounds; a coarse, inferior sand, 8 lbs., and an almost worthless heath, (gravelly soil), gained but 3 pounds

The capacity of soils for retaining water, is somewhat proportionate to that of absorbing it:—

				Of its own weight.	
Quartz sand is saturated when it contains				24	per cent.
Calcareous sand	"	"	"	28	"
Loamy soil	"	"	"	38	"
Clay loam	"	"	"	47	"
Peat (about)	"	"	"	80	"

It is thus evident, that perfection is not obtained in either sandy, gravelly, clay, or peat soils, as they are characterized in the classification we have assumed. It is only when they have been improved by partial admixture with each other, and charged with the proper quantity of vegetable manures, and the salts which are requisite for their fertility; when they have been drained, wherever necessary to free them from stagnant water, whether upon or within the soil, or to remove any noxious springs, which sometimes contain matters in solution, injurious to vegetation; and finally, when the subsoil is in the proper condition to facilitate the free admission and escape of moisture and air, and the extension of the roots in every direction—it is only when all these conditions exist, that the fullest products from soils can be realized.

It is absolutely essential to profitable cultivation, that all the earthy substances required by the crops, should exist in the soil in sufficient quantities, and in an accessible form, to supply their wants. The proportions may be various, one sometimes greatly predominating over another, as is sufficiently obvious in the equally-productive powers of good clays, sands, and peats; yet in every instance it will be found, unless owing to a heavy coating of manures, and a peculiarly favorable season, that a soil can be relied on for such constant results, only when it has been so ameliorated as to approximate towards the character of loams.

The following is an analysis of three specimens of very fertile soils, made by Sprengel: —

	Soil near Osterbruch.	From the banks of the Weser near Hoya.	near Weserbe.
Silica. Quartz, Sand, and Silicates	84.510	71.849	83.318
Alumina	6.435	9.350	3.085
Oxides of Iron	2.395	5.410	5.840
Oxides of Manganese	0.450	0.925	0.620
Lime	0.740	0.987	0.720
Magnesia	0.525	0.245	0.120
Potash and Soda extracted by water	0.009	0.007	0.005
Phosphoric Acid	0 120	0.131	0.065
Sulphuric Acid	0.046	0.174	0.025
Chlorine in common Salt	0.006	0.002	0.006
Humic Acid	0 780	1.270	0.800
Insoluble Humus	2.995	.550	4.126
Organic matters containing Nitrogen	0 960	2.000	1.220
Water	0.029	0.100	0.150
	100.	100.	100.

The above had remained for a long time in pasture, and the second was remarkable for the fattening qualities of its grass when fed to cattle.

The following are arable lands of great fertility:

	1	2		3
		From Ohio.		
	Soil from Moravia.	Soil.	Subsoil.	Soil From Belgium
Silica and fine Sand	77.209	87.143	94.261	64.517
Alumina	8.514	5.666	1.376	4.810
Oxides of Iron	6.592	2.220	2 336	8.316
Oxide of Manganese	1.520	0.360	1.200	0.800
Lime	0.927	0.564	0.243	Carb. of Lime. 9.403
Magnesia	1.160	0.312	0.310	Carb. of Mag. 10.36
Potash, chiefly combined with Silica	0.140	0.120 }	0.240	{ 0.100
Soda, ditto	0.640	0.025 }		} 0.013
Phosphoric Acid combined with Lime and Oxide of Iron	0.651	0.060	trace	1.221
Sulphuric Acid in Gypsum	0.011	0.027	0.034	0.009
Chlorine in common Salt	0.010	0.036	trace	0.003
Carbonic Acid united to the Lime	—	0.080	—	—
Humic Acid	0.978	1.304	—	0.447
Insoluble Humus	0.540	1.072	—	—
Organic substances contain'g nitrogen	1.108	1.011	—	—
	100.	100.	100.	100.

" Of these soils, the first had been cropped for 160 years successively, without either manure or naked fallow, The second was a virgin soil, and celebrated for its fertility. The third had been unmanured for twelve years, during the last nine of which it had been cropped with beans, barley, potatoes, winter barley and red clover, clover, winter barley wheat, oats, naked fallow."—(*Johnston.*)

Bergman found that one of the most fertile soils in Sweden centained 30 per cent. of carbonate of lime. Chaptal analyzed a very productive soil in France, which gave near 25 per cent of the same, and seven of organic matter. Tillet even found one, and that the most fertile, which yielded 37.5 of carbonate of lime. Some of the best in the Mississippi valley, have yielded upon analysis, 20 per cent. of magnesian lime ; and of phosphate of lime, two to three per cent. Many other soils throughout the United States, contain an equal proportion of carbonate of lime. Such are always the last to wear out, and the first to recover by the addition of manures, when suffered to remain uncultivated, or in a state of rest

CHAPTER II.

MANURES.

WHILE soils are permitted to remain in their natural state, or if denuded of their original foliage and used only for pasture, little or no change is perceptible either in their character or productive powers. A slight change, however, is gradually wrought in their texture and capacity for production, which is fully revealed in the lapse of centuries. The elevated mountain's side, and the steep declivities of hills, support an annual vegetation of more or less luxuriance; and a portion of this, together with the broken twigs and the wasting trunks of fallen trees, are carried down by the rains, and become a rich addition to the lower soils on which they ultimately rest. Beside the vegetable matter thus annually removed from one spot and accumulated upon another, many of the fertilizing salts, which the action of the roots, or exposure to the atmosphere has rendered soluble, and the finer particles of earth, which the alternations of heat and frost, of rain and drought, have reduced to dust, are also washed out of the higher soils and deposited on the plains and valleys below. Such doubtless, was once the condition of those secondary bottom-lands, which for ages, probably, received the rich deposits from other soils; but whose present situations, elevated beyond even the extraordinary rise of the rivers whose course is near, show some radical alteration of their respective levels, by which the inundations no longer contribute to their fertilization.

These soils being well stored with the food of plants, and frequently to a great depth, will bear large successive crops for a long period: and they have in many instances, been treated by their first occupants as if they were inexhaustible. Of this description, were the James River and other alluvial lands in Virginia, some of which were continued in uninterrupted crops of corn and tobacco for more than a century, without the addition of manures. But they have long since become exhausted; and the more careful planters are now endeavoring to resuscitate those worn-out lands, which ought never to have become impoverished. Of the same character are most of the secondary bottoms on the Connecticut, the Scioto, the Miami, and other rivers. The first. although under cultivation for more than two centuries, has fully maintained its productiveness, the necessary result of its minute subdivisions among intelligent farmers; and the two latter, if properly managed, are capable of perpetual fertility. Although but a little more than half a century has elapsed since these last have been subject to the white man, they have, in too many instances, already been severely cropped. The writer has seen fields, which he was assured have yielded forty-seven large successive crops of corn, and exclusively from their own resources. A more careful tillage, however, is now becoming general.

The lower alluvial bottoms that are frequently overflowed, and thus receive large coatings of manures, which are fully equivalent to the products taken off, are the only soils which will permanently sustain heavy crops without the aid of man. Such are the banks of the Nile and the Ganges, and many of our own rivers, which by the overflowing of their waters alone, have continued to yield large annual burthens; the two former, probably for more than 4,000 years; but they are thus supported, at the expense of a natural drainage of thousands of acres, which by this means, are proportionally impoverished. Manures, then, in some form, must be considered as absolutely essential to sustaining soils subjected to tillage.

In their broadest sense, manures embrace every material, which if added to the soil, tends to its fertilization. They are appropriately divided into *organic and inorganic;* the first embracing animal and vegetable substances, which have an appreciable quantity of nitrogen; the last, comprehending only such as are purely mineral or earthy, and which in general, contain no nitrogen. These characteristics are sometimes partially blended, but they are sufficiently distinct for general classification.

Much pertinacity has been exhibited by some highly intelligent minds, who should have entertained more liberal views, as to the peculiar kinds of manures necessary to support a satisfactory productiveness. We have seen that Tull maintained, that the deepening and thorough pulverization of the soil was alone sufficient to secure perpetual fertility. But this crude notion, it is evident to the most super

2*

ficial modern reader, is wholly untenable. Some agriculturists of the present day, however, while they scout Tull's theory, (who was, nevertheless, a very shrewd man for his time), will yet claim as essential to successful vegetation, the existence in the soil, of but a part only of the food of plants. Thus, one asserts that the salts alone will secure good crops ; others maintain that the nitrogenous substances are the true source of fertility; while still another class refer to the presence of humus or geine, (the available product of vegetable and animal decay in the soil), as the only valuable foundation of vegetable nutriment in all manures. Truth and sound practice lie between, or rather in the combination of all these opinions.

It has been shown in a preceding page (17th), that all fertile soils must have not less than fifteen, and more probably sixteen, different simple or elementary substances, in various combinations with each other. All of the ordinary cultivated plants, contain *potash, soda, lime, magnesia, alumina, silica, oxide of iron, oxide of manganese, sulphuric acid, phosphoric acid, chlorine,* and frequently *iodine ;* each of which, excepting the two last, are in combination with oxygen. In addition to these, they also have carbon, oxygen, nitrogen and hydrogen. Other substances or ultimate principles may possibly exist in plants, which analysis may hereafter detect, but which have hitherto eluded the closest investigation.

It is therefore obvious, that such principles as all fertile soils furnish to vegetables, must be contained in manures. It is no satisfactory answer to this position, to assert, that numerous experiments have apparently been successful, of growing plants in pure sand and water; or with charcoal and the salts added ; or even that there are some atmospheric plants, that fulfill their zoophytic existence in air. Growth may continue for a long time under such circumstances ; *but full maturity never arrives, and probably never can, without the available presence in the soil, of every element which enters into the composition of plants.*

Profitable farming requires, that manures embodying all these elements, should be added in sufficient quantities to the soil, to develope fully and rapidly, such crops as are sought from it. It becomes then, a matter of the highest consequence to the farmer, to understand not only what substances may be useful as manures ; but also, how to apply them in the best mannei to his crops, so far as they can be made profitable. We shall fu st speak of the inorganic manures.

ASHES.

If any organic matter, whether animal or vegetable, be burned, an incombustible substance remains behind, called the ash or ashes. This varies in different plants, from less than one, to over twelve per cent. of their whole weight. It also varies with the different soils upon which they are found, with the different parts of the same plant, and in the different stages of its maturity. Thus, plants which grow on peaty or low wet soils, give a less proportion of ashes than those which mature upon soils that are dry, or rich in the silicates and salts. The bark, leaves and twigs, give much more of ash than the trunks of trees and stems of plants; and in their early growth, they yield a much larger proportion than after they have attained maturity.

The following table, constructed from several reliable sources, but principally from Sprengel, arranged in part by Johnston, will show the relative quantity of ashes found in some of the more important objects of cultivation.

	Potash.	Soda	Lime.	Magnesia.	Alumina.	Silica.	Sulphuric Acid.	Phosphoric Acid.	Chlorine.	Oxide of Iron.	Oxide of Manganese.	Total in every 1000 lbs.
Wheat—Gr'n	2.25	2.40	0.96	0.90	0.26	4.00	0.50	0.40	0.10	trace		11.77
" St'w	0.20	0.29	2 40	0.32	0.90	28.70	0.37	1.70	0.30			35.18
Barley—Gr'n	2.78	2.90	1.06	1.80	0.25	11.82	.59	2.10	0.19	trace		23.49
" St'w	1.80	0.48	5.54	0.76	1.46	38.56	1.18	1.60	0.70	0 14	0.20	52.42
Oats — Grain	1.50	1.32	0.86	0.67	0.14	19.76	0 35	0.70	0 10	0.40		25.80
" Straw	8.70	0.02	1.52	0.22	0.06	45.88	0.79	0.12	0.05	0.02	0.02	57.40
Rye — Grain	5.32	*	1.22	0.44	0.24	1.64	0.23	0.46	0.09	0 42	0.34	10.40
" Straw	0.32	0.11	1.78	0.12	0.26	22.97	1.70	0.51	0.17			27.93
Field) Bean	4.15	8.16	1.65	1.58	0.34	1.26	0 89	2.92	0.41			21.36
Bean } Straw	16.56	0.50	6.24	2.09	0.10	2.20	0.34	2.26	0.80	0.07	0.05	31.21
Field) Pea	8.10	7.39	0.58	1.36	0.20	4.10	0.53	1.90	0.38	0.10		24.64
Pea } Straw	2.35		27.30	3.42	0.60	9.96	3.37	2.40	0.04	0.20	0.07	49.71
Pota-) Roots	4.028	2.334	.331	.324	.050	.084	.540	.401	.160	.032		8.284
toes } Tops	8.19	.09	12.97	1.70	.04	4.94	.42	1.97	.50	.02		30 84
Tur-) Roots	2.386	1.048	.752	.254	.036	.388	.801	.367	.239	.032		6 303
neps } Leav's	3.23	2.22	6.20	.59	.03	1.28	2 52	.98	.87	.17		18.09
Carrots	3.533	.922	.657	.384	.039	.137	.270	514	.070	.033	.060	6.619
Parsneps	2.079	.702	.468	.270	024	.162	.192	100	.178	.005	?	4.180
Rye Grass	8.81	3.94	7.34	0.90	0 31	27.72	3.53	0.25	0.06			52.86
Red Clover	19.95	5.29	27.80	3.33	0.14	3.61	4.47	6.57	3.62			74.78
White Clover	31.05	5.79	23.48	3.05	1 90	14.73	3.53	5.05	2 11	0.63		91.32
Lucern	13.40	6.15	48.31	3.48	0 30	3.30	4.04	13.07	3.18	0.30		95.52
Sainfoin	20.57	4 37	21.95	2.88	0.66	5.00	3.41	9.16	1.57			69.57

In the foregoing table, the grain, beans, peas, straw and

* Included in Potash

hay are estimated after they have been dried in the air and the roots as they are taken from the field. The clovers and grass lose from 55 to 75 per cent. of their entire weight when full of sap, lessening of course, as they approach to the state of ripening their seed. The potato loses in drying, 69 per cent. of water; the turnep, 91; the carrot, 87; the turnep leaf, 86; the carrot leaf, parsnep and parsnep leaf each 81; and the cabbage, 93.

There is much variation in the different specimens of the above substances subjected to examination, according to the peculiar variety, the different circumstances and various stages of their growth. The oat is the most variable of the grains, one specimen sometimes containing three times the quantity of ash afforded by others. The roots also, sometimes vary as three to one in their quantity of ash. As the grain and most of the other crops approach maturity, the quantity of some of the principles constituting the ash, lessens, as of potash and soda, their presence being no longer necessary in the sap, to aid the formation of the various products of the plants.

Later and probably more accurate analyses, give considerable variations in the relative quantities of the elements of the ash of different plants. Thus, an average of six of these, gave of peas with the pod, about 35 per cent. of phosphoric acid, and of beans, about 32 per cent. But the table is given to illustrate principles in the organization of plants, rather than to define the precise relative proportions of the constituents of each.

The farmer will perceive from this table, the great value of ashes to his crops. The quantity seems small in comparison with the total weight of the vegetable; yet small as it is, the aggregate of a few years will so far exhaust the soil of one or more of the principles necessary to sustain a luxuriant vegetation, that it will cease to yield remunerating returns. The annual exhaustion of salts from large crops of grain, roots and grass, is from 180 to more than 250 lbs. in every acre of soil. The ashes of vegetables, consist of such elements as are always required for their perfect maturity, and it is evident, they must furnish one of the best saline manures, which can be supplied for their growth. They are to the earthy parts of vegetables, what milk is to the animal system, or barn-yard manures to the entire crop; they contain every element, and generally in the right proportions, for insuring a full and rapid growth.

Ashes are also among the most economical manures, as from the free use of fuel in the United States, they are produced by almost every household. Good husbandry dictates, that not a pound of ashes should be wasted, but all should be saved and applied to the land; and where they can be procured at a reasonable price, they should be purchased for manure. Leached ashes, though less valuable, contain all the elements of the unleached having been deprived only of a part of their potash and soda. They may be drilled into the soil with roots and grain, sown broadcast on meadows or pastures, or mixed with the muck-heap. They improve all soils not already saturated with the principles which they contain.

The quantity of Ashes that should be applied to the acre, must depend on the soil and the crops cultivated. Potatoes, turneps and all roots; clover, lucern, peas, beans, grain and the grasses are great exhausters of the salts, and they are consequently much benefitted by ashes. They are used with decided advantage for the above crops in connection with bone-dust; and for clover, peas and roots, their effects are much enhanced when mixed with gypsum. Light soils should have a smaller, and rich lands or clays, a heavier dressing. From 20 to 30 bushels per acre for the former, and 50 for the latter, is a moderate application; or if they are leached, the quantity may be doubled with decided benefit, as they act with less energy. Repeated dressings of ashes, like those of lime and gypsum, without a corresponding addition of vegetable or barn-yard manures, will eventually exhaust tillage lands of their carbonaceous and organic matters.

Ashes may be applied to meadow-lands, for a longer time than to any other crops, and for this obvious reason. The whole surface of the soil is closely covered with vegetable agents, which are actively employed in drawing carbon from the air and soil, a large portion of which is stored up in the stubble and roots, which thus makes it less important that the organic matters should be given back to the soil, in the shape of vegetable or animal manures. As an instance of the rapidity with which this operation goes forward, it has been found, that the dried roots and stubble of a clover-field the second year, and after one crop for the first, and two for the second season had been taken off, yielded 56 lbs. for every 100 lbs. of the aggregate crops of hay. An old meadow has yielded 400 lbs. of roots for every 100 of hay for the season.

The carbon is constantly increasing in the soil of well

managed pastures, and it also increases for a time in meadows. It will continue to do so for an indefinite period, if the ashes of plants are added to the soil, nearly to the amount of the mineral ingredients taken off. With this increase in the organic elements of vegetation, (if we were certain that nitrogen is accumulated in the same ratio, which we are not), it is evident that the salts alone would then be wanting to give the utmost luxuriance ; and these are found combined in the most convenient and generally the most economical form as ashes. But care is necessary that they be not added in excess.

Coal Ashes.—The bituminous and anthracite coals afford ashes, and although inferior in quality to those made from wood and vegetables, are like them, a valuable manure, and they should be applied to the land in a similar manner. If they contain many cinders, from not having been thoroughly burned, they are more suited to heavy than to light soils, as they tend to their mechanical division, which though beneficial to the former, are injurious to the latter.

Ashes of Sea Weeds or Marine Plants.—When from its quantity or remoteness, it is inconvenient to carry the sea-weed to the soil, which abounds on most of our sea-coasts, it can be burned ; when it will be found to yield a large proportion of ash, which is peculiarly rich in soda. This is of great value to the farmer. Several species of the *fuci* have for a long period been collected and burned on the northern coasts of Scotland, Norway and the Baltic, forming an article of commerce under the name of *kelp.* Its value consisted in its alkaline properties, for which it was much used by the glass and soap-makers, the bleachers, and for other uses in the arts. For these purposes, it is now nearly superseded by *soda ash,* a crude carbonate of soda, extracted by the decomposition of sea-salt ; and the price it now bears in market, will bring it within the reach of farmers for some of the economical purposes of husbandry.

Peat Ashes.—Nearly all peat approaching to purity, when thrown out of its bed and thoroughly dried, will admit of being burned to an imperfect ash ; and when it does not reach this point, it will become thoroughly charred, and reduced to cinders. In both of these forms, it is a valuable dressing for the soil. It is always better for dry uplands, to use the unburned peat after it has been properly composted in a muck heap ; as the organic matters which it contains, and which are expelled by burning, are of great benefit to the soil. But when they are remote, the peat may be burned at a trifling cost, and the ashes applied with manifest profit. The principal use hitherto made of them by farmers, has been in spreading them directly over the surface of the reclaimed bed from which they were taken.

LIME.

LIME is the product of limestone, marble, chalk, or marl, after it has been burned, or subjected to an intense heat. In either of the foregoing forms, it is a carbonate, and contains from 43 to 46 per cent. of its weight of carbonic acid, which is expelled by calcination. After the acid has been driven off, it exists in its quick or caustic state; and in that condition, its affinity for moisture and carbon is so great, that it greedily combines with both, on exposure to water, earth, or even to the atmosphere, passing again into a carbonate and hydrate. It is in these latter conditions, that it is applied to soils and muck heaps. If reduced to an impalpable powder, (the condition in which chalks and marls generally exist,) limestone would act with equal efficiency as if burnt.

Lime, next to ashes, either as a carbonate or sulphate, has been instrumental in the improvement of our soils, beyond any other saline manures. Like ashes, too, its application is beneficial to every soil, not already sufficiently charged with it. It makes heavy land lighter, and light land heavier; it gives adhesiveness to creeping sands or leachy gravel, and comparative openness and porosity to tenacious clays; and it has a permanently beneficial effect, where generally used, in disinfecting the atmosphere of any noxious vapors existing in it. It not only condenses and retains the volatile gases brought into contact with it by the air and rains, but it has the further effect, of converting the insoluble matters in the soil, into available food for plants. It has proved, in many instances, the wand of Midas, changing everything it touched into gold. It is the key to the strong box of the farmer, securely locking up his treasure till demanded for his own use, and yielding it profusely to his demands whenever required. In its influence in drying the land, and accelerating the growth of plants, the use of lime is equivalent to an increase of temperature; and the farmer sometimes experiences in effect, the same benefit from it, as if his land were removed a degree or two to the south. The influence of lime in resuscitating soils after they have been exhausted, has been frequent and striking; and it may be stated as an incontrovertible truth, that wherever procurable at low prices, lime is one of the most economical and efficient agents in securing fertility, within the farmer's reach.

It has been falsely said to be an exhauster of soils; that it enriches the fathers and impoverishes the sons. So far as it gives the occupant of the land the control over its latent fertility, this is true; but if he squanders the rich products when within his reach, it will be his own fault. Lime gives him the power of exhausting his principal; if he uses aught beyond the interest, his prodigality is chargeable to his own folly, not to the liberality of his agent. By the addition of lime to the soil, the insoluble ingredients contained in it are set free, and they are thus enabled to aid in the formation of plants; and larger crops, and of better quality, are the results. If these be taken from the soil without a corresponding return of manure, exhaustion must follow. In the preceding table, it is seen, that lime constitutes, in all cases, only a very minute part of the entire plant; all the other ingredients must be added, or the fertility of the soil cannot be sustained. But in the very abundance of the crops which lime affords, means are provided for the maintainence of the highest fertility. If they are consumed on the farm, their manure should be returned to the fields; and if sold, other manures should be procured to replace the substances from which they are formed.

A practice which has extensively prevailed for many years, in sections of the eastern states, consists in alternating wheat and clover on strongly limed-lands. The plan usually adopted, is to give one year to wheat, and one or two following, to clover; sometimes taking off the first clover crop for hay, and feeding off upon the ground, and plowing in, the after growth for manure; and upon this, wheat is again sown. This course has succeeded in bringing into fine condition, many unprofitable fields. It may work well for years, but it is nevertheless faulty and improvident. Lime only is added directly to the soil, but the clover draws from the air and moisture, whatever food it can attract from them. There remain to be supplied, potash, soda, the phosphates and silicates, (which the soil will soon cease to furnish sufficient for the wants of the wheat and clover removed), or sterility must inevitably follow.

The best method *is to add in some form, the full amount of all the material abstracted by the annual crop*

When this is done, the large dressing of lime will retain the accumulating fertility, far beyond what the soil would be capable of were it not for its agency; and it is in this that the great profit of farming consists.

Large crops only are profitable. The market value of many indifferent ones, will hardly meet the expense of cultivation, and it is only the excess beyond this, which is profit. It is evident that if 15 bushels per acre of wheat, be an average crop, and it requires 12 bushels to pay all expenses of production, three bushels is the amount of profit. But if by the use of lime and ordinary manures, the product can be raised to 30 bushels per acre, the profit would be near the value of 12 or 15 bushels, after paying for the manures. Thus the advantage from good management, may be five times that of neglect. This example is given as illustrating a principle, and not as an exact measure of the difference between limed and unlimed fields.

Application of Lime.—Lime may be carried on to the ground immediately after burning, and placed in small heaps. There it may be left to slack by rains and the air; after which spread it preparatory to plowing. A good practice is to place it in large piles, and cover it thickly with earth, which gradually reduces it to powder. It may then be carried where it is wanted, and spread from the cart. It is still better, when small quantities only are wanted, to add it to the compost, after it has been thoroughly air-slacked, avoiding fermentation, as far as practicable, after it has been added; as its avidity for carbon expels the ammonia, which is the most valuable of the volatile ingredients of the muck heap. A thick coating of earth over the whole, will arrest and retain much of the gas that would otherwise escape. Fresh burnt lime does not act beneficially upon the crops during the first year of its application, unless prepared by adding it to three or four times its bulk of earth, in which condition it should remain for a long time. If it be mixed with rich mold when first taken hot from the kiln, it will decompose or liberate the alkalies contained in the earth to such a degree, as to render this compost a powerful manure.

Nearly all limestones yield lime sufficiently pure for agricultural purposes. When required of greater purity, it may always be obtained by burning oyster shells, or others of marine origin.

Magnesian Lime.—Many of the limestones contain magnesia, and are called magnesian lime. The effect of this, is a more energet c action, and where it is found in lime, the same result wi.l be produced by the application of a less quantity.

The amount of Lime to be used, depends entirely on the soil. Some fertile lands contain over 30 per cent. in their natural state. The large amount, of more than 600 bushels of lime per acre, has been applied at one time, to heavy clays, and such soils as were full of vegetable mold, with de- cided benefit to the land. But equally beneficial results would have been produced, had one half the quantity been first added, and 50 bushels every third or fourth year subse- quently. In the United States, the average for a first dress- ing, is from 50 to 120 bushels per acre. This may be re- newed every four or five years, at the rate of 20 to 40 bush- els. If an overdose has been applied, time, or the addition of putrescent or green manures, are the only correctives.

To give lime its fullest effect, it should be kept as near the surface as possible; and for this reason, it is well to spread it after plowing, taking care to harrow it well in. Allow it then to remain in grass as long as profitable. Its weight and minuteness give it a tendency to sink; and after a few years of cultivation, a large portion of it will be found to have got beyond the depth of its most efficient action. Where lime is used, this tendency gives additional value to the system of underdraining and subsoil plowing, which enables the atmosphere and roots to follow it, thus prolonging its effect, and greatly augmenting the benefit to crops. It should be spread upon the ground immediately after taking off the last crop, so as to allow the longest time for its action before the next planting.

Application to Meadows.—In addition to its other good effects, lime, like ashes, is useful to meadows in destroying the mosses, and decomposing the accumulated vegetable matters on the surface. For this purpose, it may be spread on them unmixed, after having first passed into the state of carbonate or effete lime, to prevent injury to the grass. If no such necessity require its use in this form, it may be com- bined advantageously with the muck and scattered broad cast over the meadow.

MARLS.

Marls are composed of carbonate of lime, mixed with clay, sand, or loam, and frequently with sulphate and phosphate of lime They are a useful application to land in proportion to the lime they yield ; and when containing the phosphate in addition, their value is largely increased. The quantity that may be advantageously used, is even more variable than that of pure lime, inasmuch as the quality varies with every bed in which it is found. They are adapted to the improvement of all soils, unless such as are already sufficiently filled with lime, and they are more generally useful to meadows than the pure carbonate. Their benefits will be greatly enhanced, if the clay marl be used on light or sandy soils, and sandy marls on clays and heavy lands. Marl has sometimes been applied at the rate of 200 cubic yards per acre; but where it approaches to purity, and the soil is in the proper condition to be benefitted, even four or five cubic yards may be sufficient to produce the best effects. Circumstances alone must determine the proper quantity to be used. Marl should be carried out and exposed in small heaps, before spreading on the land. Exposure to the sun, and especially to the frosts of winter, is necessary to prepare it for use.

Analysis of Marls.—Marls may be readily analyzed by any one, with a pair of accurate scales and weights and a large-mouthed vial. To one part muriatic acid, add two parts water, fill the vial to about one third, and balance it on the scales. Then slowly add 100 grains of the pulverized marl, thoroughly dried over the fire. When the effervescence has subsided, expel the carbonic acid from the vial, by pouring off, or blowing into the vial through a reed or with a bellows, its greater weight causing it to retain its place to the exclusion of the air. Now add weights to the opposite scale till balanced, and the deficiency of grains under 100, will show the amount of carbonic acid expelled; and as this is combined in the proportion of 46 to 54 of *quick* or pure lime, in every 100, the loss indicates 46 per cent. of the carbonate of lime contained in the marl.

From the frequent presence of phosphate and sulphate of lime, and sometimes potash and animalized matters in marls, this kind of analysis seldom indicates the value of a marl bed for agricultural purposes. If its exact worth is to be ascertained, there must be a more perfect analysis, by an experienced chemist.

SHELL SAND.

This is a calcareous sand, sometimes mixed with animal matter and the phosphates. It abounds on some parts of the coast of Cornwall, and on the western shores of Scotland.

and Ireland. It is also found on the coast of France, and particularly in Brittany, where it is known by the name of *trez*. This produces prodigious effects on peaty, clay and other soils, to which it is applied at the rate of 10 to 15 tons per acre. It is so much esteemed for the former, that it is sometimes carried to a distance of 100 miles. It is probable, there are similar deposits on the coast of some of the Atlantic States, though I am not aware of its application for agricultural objects. Its great value as a top dressing, will fully justify exploration, for the purpose of detecting it wherever it may exist.

GREEN SAND MARL.

There are extensive beds of a green sand, (generally though improperly termed marl), which run through a section of New Jersey, from which farmers have derived an astonishing addition to their crops. Much of it is found by analysis, to contain scarcely an appreciable quantity of lime, but it readily yields a large amount of potash, varying from six to 15 per cent. From a careful analysis of eight different specimens, Professor Rodgers found in it an average of 10 per cent. of potassa. The effect of this applied to the barren sands, which abound in that neighborhood, has been so favorable, that lands which before could be bought for three dollars per acre, would afterwards bring $40. Several deposits of green sand in the counties of Plymouth and Barnstable, Mass., similar in external appearance to the foregoing, were explored by Professor Hitchcock, and specimens were analyzed by Dr. Dana, without, however, detecting any qualities of decided advantage to agriculture.

GYPSUM—PLASTER OF PARIS—OR SULPHATE OF LIME.

This is a combination of lime with sulphuric acid and water, in the proportion of 28 of lime, 40 of acid, and 18 of water. It is frequently found in connection with carbonate of lime, clay, &c. The use of gypsum has been attended with great benefit in most parts of the United States; and by many of the most experienced farmers in certain localities, it is justly considered as indispensable to good farming. Like other saline, and indeed all manures, it acts beneficially only on soils which are naturally dry, or have been made so by artificial drainage. It is felt most on sands and loams; but generally, it is advantageously added to clay soils, requiring more for the latter, and for all such as contain a large proportion of vegetable matter. From two pecks on sandy, to fifteen bushels on clay soil, have been applied per acre ; but from two to four bushels is the usual quantity.

The crops on which it produces the greatest effect, are the red and white clover, lucern and sainfoin, and the legumi-nous plants, peas, beans, &c. On natural meadows and the cereal grains, it has little perceptible influence.

Gypsum should be sown broadcast on meadows, as soon as the first leaves have expanded in the spring. It requires 460 times its weight of water to dissolve it, which shows the ne-cessity of applying it while the early rains are abundant. For corn, potatoes and turneps, it is usually put in with the seed, or sprinkled upon the leaves after the first hoeing ; and it is advantageously applied in both ways, during the same season.

From its great effect on the clovers, increasing them sometimes to twice, and in rare instances, to thrice the quan-tity produced without it, it is manifest that gypsum is the most profitable manure which can be used, as it can be gen-erally procured by farmers at from $3 to $6 per ton. Yet it should be fully understood, that like lime, salt, or other mineral manures, it furnishes a part only of the food of plants ; and like them too, the addition of vegetable and ani-mal manures, is indispensable to secure permanent fertility.

Some sections of this and other countries, particularly in Great Britain, apparently derive no benefit from the applica-tion of gypsum. This failure has been variously ascribed, to there being already enough in the soil, or to the presence of a marine atmosphere. Its great usefulness, however, on many parts of our Atlantic coast, would seem to require some other explanation than the last, as the cause of its in-efficiency. Experiments alone can determine the circum-stances which will justify its application, and to this test should not only this, but all other practices of the farmer be rigidly subjected.

BONES.

About 33 per cent. of fresh bone, consists of animal mat-ter, (oil, gelatine, &c.,) from 53 to 56 per cent. of phosphate of lime, and the remainder is principally carbonate of lime, soda and magnesia. There is no part of the bone that is not useful to vegetation ; and it is especially so to the various kinds of grain, potatoes, turneps, the clovers, peas and beans. The bones should be crushed or ground, and then drilled in with the seed, or scattered broadcast, at the rate of

25 bushels per acre. They may be repeated in less quanti
ties every four or five years, or till the soil ceases to be improved by them, when they should be withheld till additional cropping shall have so far exhausted them, as to justify a
further application.

Bones are generally boiled before using for manure, to
extract the oil and glue. This does not lessen their value
for agricultural purposes, beyond the diminution of their
weight, while it hastens their action. They are sometimes
burned, which drives off all the organic matter, leaving only
the lime and other bases, to benefit the soil. This is a
wasteful practice, though the effect is more immediate on the
crops ; but it is also more transient, and they require to be
more frequently renewed.

Bones ought always to be saved ; and if not practicable
to crush them, they may be thrown upon the land, where
they will gradually corrode and impart their fertilizing properties. When partially decomposed and buried just beneath
the surface, the roots of the luxuriant plants above, will
twine around them in all directions, to suck out the rich food
which ministers so freely to their growth. Crushed bones
are advantageously used with nearly an equal amount of
ashes, or with one third their weight of gypsum ; or they
may very properly, be added to the muck heap, where decomposition will be hastened, and they will sooner be prepared to impart all their fertility to the crops.

The effect of ground bones is greatly hastened and augmented by dissolving them in sulphuric acid. This is done
by placing the bones in tight casks or large kettles, and then
adding one third their weight of acid, diluted with half its
weight of water. This mixture immediately raises the temperature to 300° Far., and decomposition of the bones soon
results. Then add to this product, two or three times its
bulk of light mold, and when dried, it may be drilled in with
the seed by a machine, or scattered broadcast. Or a heap
of fine mold may be formed like a basin, the bones thrown
into the centre and the diluted acid added ; when dissolved
all may be mixed together. The whole nutritive matter
contained in the bones, is thus set free for the use of the
plants, which would otherwise require years to accomplish,
from the great insolubility of bones. Less than one fourth
the quantity usually applied, will thus be equally beneficial
for the first season, though the prolonged effects of the larger
quantity will be greater.

Bones may be dissolved rapidly, by throwing into a compact heap, moistening, and then covering them with earth. In this condition they soon heat and crumble, and when thus reduced, they may be applied to the land.

PHOSPHATE OF LIME.

This exists in a fossil state, and is known in some of its forms as *apatite, phosphorite,* &c. An extensive quarry is found in Estramadura, Spain; large beds of it exist in various parts of England and on the Continent of Europe, associated with carbonate and other forms of lime; and small deposites of it have been discovered in various parts of the United States. It is probable, it may yet be found in such localities and in such abundance, as to be useful to the farmer. It has been shown, that more than half of the whole weight of bones consists of pure phosphate of lime; its value for agricultural purposes is therefore apparent, since the principal benefit of bones is derived from the large proportion of phosphate they contain.

SALT—OR CHLORIDE OF SODIUM

Is variously obtained, either as fossil or rock salt; from boiling or evaporation of salt springs; and from the waters of the ocean. In a pure state, it consists of 60 of chlorine and 40 of sodium, in every 100 parts. Sodium chemically combined with oxygen, forms soda; and it will be seen by referring to the table, (page 35), that salt furnishes two of the important constituents in the ash of every vegetable. Its advantages to vegetation are to be inferred from a knowledge of its composition, and this inference is fully corroborated by experience. The merits of salt as a manure, were understood and appreciated by the ancients, and by them it was extensively used on their fields. It has continued to be employed for the same purposes, by intelligent agriculturists to the present time. On some soils, it yields no apparent benefit. Such as are near the sea-coast, and occasionally receive deposits from the salt spray, which is often carried far inland by the ocean storms; or such as contain chlorine and soda in any other form, and in sufficient abundance for the wants of the crops, are not affected by it. But in other situations, when used at the rate of three to sixteen bushels per acre, the crops of grains, roots or grasses have been increased from 20 to 50 per cent. It may be applied in minute portions in the hill, or scattered broadcast, or

mixed with the muck heap. Its great affinity for water, has
the effect, like that of gypsum, of attracting dews and atmospheric vapor to the growing vegetation, but in a still
greater degree. By this means, a copious supply of moisture
is secured to those plants which have been thus manured,
and far beyond what is experienced in adjoining fields. Salt
is also useful in destroying slugs, worms, and larvæ, which
frequently do much injury to the crops.

SULPHATE OF SODA, (Glauber Salts,) SULPHATE OF MAGNESIA, (Epsom Salts,) AND SULPHATE OF POTASH.

These are all useful manures, and they act on vegetation
in a manner similar to gypsum. This was to have been expected, so far as the sulphuric acid is concerned, which is
common to each; but their action is modified to a certain
degree, by the influence which the bases or alkaline ingredients of these several salts exert upon the plants. The gene-
rally-increased price which they bear over gypsum, will prevent their use, when remote from those localities where they
exist in a state of nature; or where they may be procured
at low rates, near the laboratories in which they are manufactured.

NITRATE OF POTASH, (Saltpetre,) AND NITRATE OF SODA.

These are both extensively found in a crude state in native beds, or as an efflorescence; and in this condition, they
can frequently be bought at a price that will justify their
use. The first contains potash, 46.5, and nitric acid, 53.5; the
second, in its dry state, soda, 36.5, and nitric acid, 63.5, in
every 100 parts. Numerous experiments have been tried
with them on various crops; but they have not thus far, afforded very accurate or satisfactory results. In general, they
give a darker color and more rapid growth, and they increase the weight of clover, the grasses and the straw of
grain; and the forage is also more relished by cattle. But
in the average effects upon grain and roots, the statements
are too much at variance to deduce any well-settled principles, which we might safely assume as a reliable guide to the
practical agriculturist. From the decidedly beneficial effects produced in numerous instances, may we not reasonably infer, that they have generally been successful where
there has been a deficiency of them in the soil?

As a soak or steep for seeds, and especially when dissolved and added to the bed where they are planted, there
is no doubt of their possessing some value in giving an early

and vigorous growth to most seeds, besides securing them from the depredation of insects and marauders of various kinds. This enables them rapidly to push forward their roots, stems and leaves; thus obtaining a greater range for the roots, and more mouths for the leaves to draw their nourishment from the atmosphere, by which vegetation is accelerated, and where the period for maturity is limited, materially increasing the product.

CARBONATES, NITRATES, SULPHATES, PHOSPHATES, SILICATES, AND CHLORIDES.

Several of these have just been particularly enumerated. The remainder are composed of carbonic, nitric, sulphuric and phosphoric acids, silica and chlorine, in chemical combination with potash, soda, lime and the other bases or ash of plants. Although no one of these can fail to benefit crops, when rightly applied, yet the expense of most of them, will prevent their extended use. This can only be looked for, from those which are procurable at a cheap rate. The chemical laboratories, glass works, and some other manufactories, afford in their refuse materials, more or less of these mineral manures, which would well repay the farmer for removing and applying to his land. The most obvious that occur in this country, are all that will be here mentioned.

OLD PLASTER.

This is a true silicate of lime; being formed mostly of siliceous sand and lime, chemically combined. For meadows, and for most other crops, especially on clays and loams, this is worth twice its weight in hay; as it will produce a large growth of grass for years in succession, and without other manure. This effect is due, not only to the lime and sand, but to the nitric acid which they have abstracted from the atmosphere, and which they continue alternately (while in combination) to absorb from the air and give out to the growing plant. But the farmer cannot too carefully remember, that with this, as with all other saline manures, but a part of the ingredients only is thus supplied to vegetables; and without the addition of the others, the soil will sooner or later become exhausted.

BROKEN BRICK AND BURNT CLAY.

These are composed mostly of silicate of alumina, but they are generally mixed with a small quantity of silicate of pot

3

ash and other substances. They are of much value as a top-dressing for meadows. In addition to their furnishing in themselves, a minute quantity of the food of plants, like old plaster, they serve a much more extended purpose, by condensing ammonia, nitric and carbonic acid, which they give up to the demands of vegetation. They seem to fulfil the same part as conductors, between the nutritive gases that abound in the atmosphere and the vegetables which they nourish, as the lightning rods in leading the electricity from the clouds to the earth.

CHARCOAL.

When charcoal is scattered over the ground, it produces the same effect as the foregoing, and probably in a greater degree; as it absorbs and condenses the various gases within its pores, to the amount of from 20 to over 80 times its own bulk. The economy and benefit of such applications, can be readily understood, as they are continually gleaning these floating materials from the air, and storing them up as food for plants. Charcoal as well as lime, often checks rust in wheat, and mildew in other crops; and in all cases, mitigates their ravages, where it does not wholly prevent them.

BROKEN GLASS

Is a silicate of potash or soda, according as either of these alkalies are used in its manufacture. Silicate of potash, (silex and potash chemically united,) is that material in plants, which constitutes the flinty, exterior coating of the grasses, straw, cornstalks, &c.; and it is found in varying quantity in all plants. It is most abundant in the bamboos, cane, Indian corn, the stings of nettles, and the prickly spikes in burs and thistles. Some species of the marsh-grasses have these silicates so finely, yet firmly adjusted, like saw-teeth, on their outer edges, as to cut the flesh to the bone when drawn across the finger. Every farmer's boy has experienced a yet more formidable weapon, in the slivers from a cornstalk.

It is to the absence of this material in peats and such other soils, as have an undue proportion of animal or vegetable manures, that we may attribute the imperfect maturity of the grains and cultivated grasses grown upon them, causing them to crinkle and fall, from the want of adequate support to the stem, and it is to their excess in sandy and calcareous soils, that the straw is always firm and upright, whatever may be the weight of the bending ear at the top. By a deficiency of silicates, we mean, that they do not exist in a soluble form, which is the only state in which plants can seize upon and appropriate them. The efforts of some roots in procuring this indispensable food, have been so irresistible, as to have decomposed the glass vessels in which they have been grown. Before using it as a manure, the glass should be reduced to powder by grinding.

CRUSHED MICA, FELDSPAR, LAVA, THE TRAP ROCKS, &c.

Feldspar contains 66.75 of silica; 17.50, alumina; 12, potash; 1.25, lime; and 0.75, oxide of iron. *Mica* consists of silica, 46.22; alumina, 34.52; peroxide of iron, 6.04; potash, 8.22; magnesia and manganese, 2.11. Most of the *lavas* and *trap-rocks* hold large quantities of potash, lime, and other fertilizing ingredients. The last frequently form the entire soils in volcanic countries, as in Sicily, and around Mount Vesuvius in Italy, in the Azores and Sandwich Islands; and their value for grains and all cultivated plants, is seen in the luxuriance of their crops and the durability of their soils. These examples illustrate the great influence of saline manures, and their near approach to an entire independence in sustaining vegetation. Whenever they become exhausted by the severe usage they undergo, two or three years of rest enables them again to yield a remunerating crop to the improvident husbandman.

Granite, sienite, and some other rocks, yielding large proportions of potash and some lime, abound throughout the eastern portion of this country. The potash in them, is however, firmly held in an insoluble state; but if they are subjected to a strong heat, they may then be easily crushed, when they yield the potash freely by solution. In this condition, they constitute a valuable top-dressing for almost every soil and crop.

It is a subject of frequent remark, that the soil underneath, or in immediate contact with certain walls, which have been erected for a long period, is much richer than the adjoining parts of the same fields. This difference is probably due, in some measure, to the slow decomposition of important fertilizers in the stone, which are washed down by the rains, and become incorporated in the soil. The removal of stones from a fertile field, has been deprecated by many an observing farmer, as materially impairing its productiveness.

In addition to the shade afforded by them against an intense
sun and protection from cold winds, their influence in condensing moisture, and the beneficial effects which perhaps
ensue, as in *fibrous covering,* the difference may be attribut·
able to the same cause.

SPENT LYE OF THE ASHERIES

Is the liquid which remains, after the combination of the
lye and grease, in manufacturing soap.   It is of great value
for plants.   Before applying it to the land, it should be
mixed with peat or turf, or diluted with ten times its bulk of
water.   Five gallons of this lye, is estimated to contain as
much potash or soda, according as either is used, as would
be furnished by three barrels of ashes.   It has besides, a
large quantity of nitrogen, the most valuable ingredient of
animal manure, which, by judicious application, is either
converted into ammonia, or serves the same purpose in yielding nutrition to plants.

AMMONIACAL LIQUOR (from the gas houses).

This liquid is the residuum of bituminous coal and tar used
in making gas, and holds large quantities of nitrogen, from
which ammonia is frequently extracted.   When used for
land near by, it may be carried to the muck heap in barrels;
and when at remote distances, gypsum or charcoal dust may
be added to the barrel, stirring it well for some time, and
then closely covering it.   The gypsum and charcoal soon
combine with the ammonia, when the liquid may be drawn
off, and the solid contents removed.   It is a powerful manure
and should be sparingly used.

GUANO.

Guano is derived exclusively from the animal creation;
but from its existence in a highly-condensed state, and in
combination with large proportions of the salts, and having,
by its accumulation through thousands of years, lost the distinguishing characteristics of recent animal matter, it may
be almost considered as a fossil, and as properly enough
classed under the head of inorganic manures.   It is the remains of the excrements, food and carcasses of innumerable
flocks of marine birds and seals, which have made some of
the islands in the Pacific and Atlantic oceans, places of resort
for rearing their young, through unknown ages.

Peruvian Guano is found on the islands of the Pacific

near the coast of Peru and some of the headlands on the adjacent shores, between latitudes 13° and 21° South. It is here deposited to the depth, sometimes, of 50 or 60 feet. Within the degrees above named, rain seldom falls; and there is little waste, either of the substance or quality of these vast accumulations, from the lapse of time or the action of the elements.

The water-fowl which resort to this coast and the islands near it, subsist principally on fish; and their feces are consequently, much richer in nitrogen than those of any species of the feathered tribes, excepting such as are exclusively carniverous.

Peruvian guano is of a light, brown color, resembling the yellowish earths or loam; and it is beyond all comparison, superior to any other guano yet discovered, or than other manures hitherto known. The following average analysis of Dr. Ure, shows that this description of guano, contains the important and rarer portions of animal manure, in proportions far beyond that of any animal matter in its natural combinations. In every 100 parts, there are, of

Organic matter containing nitrogen, including urate of ammonia, and capable of affording from 8 to 17 per cent. of ammonia by slow change in the soil, ... 50

Water 11. Phosphate of lime 25, ... 36

Ammonia, phosphate of magnesia, phosphate of ammonia and oxalate of ammonia, containing from 4 to 9 per cent. of ammonia, ... 13

Silicious matter from the crops of birds, ... 1

Its character, as correctly indicated by such an analysis as the above, and which is fully sustained by the astonishing, and generally profitable results that follow its application has rendered it, though of recent introduction, one of the most popular manures, both in America and Europe. It has been known and appreciated by the Peruvians from time immemorial; and by its liberal use alone, combined with irrigation, they have for ages, produced the most abundant crops of Indian corn and wheat. It was scarcely known in Europe till 1840. Extensive experiments were then made with it in Great Britain. These were so satisfactory, that over 375,000 tons have since been imported into that country in a single year.

African or Ichaboe and Patagonian Guano have been brought into this country to a limited amount. They have been used with some advantage, but their value is far below that of the Peruvian.

The first introduction of Guano into this country, was in 1825. It was used in a few gardens to a limited extent, and then forgotten. Soon after its successful appearance in England, its importation was commenced in the United States. Owing to the diminished value of our agricultural products as compared with those of England, the progress of our importation has been slow; but it has been steadily advancing, and will probably reach the amount of 20,000 tons for the ensuing year (1850). It already occupies the rank of a staple import; and its constantly-increasing use hereafter, along the Atlantic and Pacific borders of the United States, may with certainty be predicted.

Guano is applied upon nearly all crops and soils; but it is, perhaps, most suited to such of the latter, as approach to sandy loams. From 200 to 500 pounds per acre, is a proper dressing, the largest quantity being required for the more sterile soils. It should be thoroughly mixed for a few days, with five times its bulk of vegetable mold or loam, and some charcoal or gypsum, after breaking the lumps and sifting in alternate layers. Avoid the use of lime or ashes in the compost, as they tend to expel the ammonia ; and keep it under cover, beyond the reach of water or rains till used. It may then be scattered broadcast, upon meadows or grain, or placed near the seeds, or young plants in the hill. A double application has been attended with the best effects; the earliest, producing a rapid and luxuriant growth of stalk, and a later one, filling out the grain, far beyond what could have otherwise been expected. The white or small grains, corn, potatoes and other roots, melons and other fruits, flowers, &c., &c., are all susceptible of the presence of guano, and are greatly benefited by it, whenever there is sufficient moisture, fully to develope its ingredients.

When used for steeps, one pound of guano is added to ten or fifteen gallons of water, then stirred well together and closely covered, (to prevent the escape of the ammonia,) for 24 hours or more, when it will be ready for use. In watering the plants, avoid sprinkling the liquid upon the stems or leaves, or it may burn or injure the plants. The surface should be freshly stirred, to admit the liquid near the roots.

As a soak for seeds previous to planting or sowing, it is frequently of great benefit in hastening germination and promoting growth ; but great care should be used that it be not made too strong.

SOOT.

Like ashes, soot has its origin exclusively from vegetables, but may, with them, be properly treated under the present head. It holds ammonia, charcoal and other important fertilizers, and is used at the rate of 50 to 200 bushels per acre. Soot produces its greatest effects in moist weather, and in dry seasons it has sometimes proved positively injurious. It may be sown broadcast over the field, and harrowed in ; or mixed with such other manures in the muck heap, as are intended for immediate use. The ammonia has a great tendency to escape, which can only be prevented by adequate absorbents, such as peat, muck, rich turf, tan bark or other vegetable remains. Many experiments made with it, have proved contradictory. In some, it has been shown to be useless for clovers ; while it has proved of great service to several of the grasses. Salt, when mixed with it, enhances its effects. In an experiment made in England with potatoes, on three separate acres of land of equal quality, one without manure gave 160 bushels ; one manured with 30 bushels of soot yielded 196 ; and the third, which received the same quantity of soot and seven bushels of salt, yielded 236. The salt insures for it that degree of moisture, which is probably essential to its most beneficial action.

CHAPTER III

~~~~~~~~~~

ORGANIC MANURES.

THE PRINCIPLES CONSTITUTING ANIMAL AND VEGETABLE-PUTRESCENT OR ORGANIC MANURES.

From the table in the foregoing pages on the ashes of plants, it is shown, that in burning dried vegetables, they lose from about 95 to 99 per cent. of their whole weight. The matter that has been expelled by heat, consists of four substances or ultimate principles ; carbon, oxygen, hydrogen and nitrogen, of which carbon makes up from 40 to 50 per cent., or about one half of the whole.

Carbon constitutes all of charcoal but the ash; nearly all of mineral coal, and plumbago or black lead ; and even the brilliant diamond is but another form of carbon. The properties and uses of carbon are various and important; its agency in the growth of plants alone, concerns us at the present time.

Carbonic Acid.—When any matter containing carbon is burned, its ultimate particles or atoms combine with the oxygen which exists in the atmosphere, and form carbonic acid, consisting by weight, of six of the former and sixteen of the latter. When animals inhale air into their lungs, a similar union takes place ; the carbon contained in the system being brought to the surface of the lungs, and after uniting with the oxygen, is expelled as carbonic acid. Pure limestone or marble loses 46 per cent. of its weight by burning ; and all of this loss is carbonic acid, which the lime slowly absorbs again on exposure to the air, or to such substances as contain it. It is evolved by fermentation ; and if the surface of a brewer's vat in full activity, be closely observed in a clear light, it may be seen falling over the edges, when it is gradually absorbed by the air. Its density is such, that it may be poured from one open vessel into another, without material loss. It is this which gives to artificial soda water and to mineral springs, (as the Saratoga,) their sparkling appearance and acid flavor. It abounds in certain caves, sunken pits, and wells, which destroy animal life, not from any intrinsic poisonous qualities, but from its excluding oxygen, which is essential to respiration. And it is from the same cause, that death ensues to such as are confined in a close room where charcoal is burnt.

This acid is an active and important agent in the incessant changes of nature. It is everywhere formed in vast quantities, by subterranean fires and volcanoes. Though heavier than atmospheric air, it mingles with it, and is carried as high as examinations have yet been made, constituting in bulk, about one part in 1,000 of the atmosphere, and something more than this in weight. Gay Lussac ascended in a balloon 21,735 feet, and there filled a bottle with air, which analysis showed to be identical in composition with that on the surface of the earth. Carbon is one of the great principles of vegetation, and it is only as carbonic acid, that it is absorbed by the roots, leaves and stems of vegetables, and by them is condensed and retained as solid matter.

Oxygen, hydrogen and nitrogen, when uncombined with other substances, exist only as gases. The first makes up nearly one half of all the substances of the globe ; and with the exception of chlorine and iodine, it constitutes a large part of every material in the ash of plants. It forms rather over 21 per cent. by measure, and 23 by weight, of the whole atmosphere ; and about eight parts out of nine, by weight, of water, hydrogen making up the remainder. It is absorbed and changed into new products by the respiration of animals, and it is an essential agent in combustion. Oxides are composed of it, in union with the metals and alkalies ; and most of the acids, when it is combined with other substances, as nitrogen, sulphur and phosphorus. Its presence, indeed, is almost universal, and the agency which it exerts in vegetable nutrition, is among the most varied and intricate manifested in vegetable life.

Hydrogen is the lightest of all the gases. It is but $\frac{1}{14}$ the weight of the atmosphere, and $\frac{1}{16}$ the weight of oxygen ; and from its great levity, it is used for filling balloons On applying a lighted taper, when brought into contact with atmospheric air, it burns with a light flame, the combustion forming water.

3*

It is largely evolved from certain springs, in connection with carbon or sulphur. This is called carburetted and sulphuretted hydrogen, an offensively pungent and inflammable gas. So abundantly is this emitted from the earth in some places, that it is used for economical purposes. The inhabitants in the village of Fredonia, New York, light their buildings with it; and some of the salt manufacturers in the valley of the Ohio, apply it to evaporating the water of the saline springs. Carburetted hydrogen is the gas now employed for lighting cities. It is manufactured from oils, fat, tar, rosin and bituminous coal, all of which yield large quantities of carbon and hydrogen. Both the carbon and hydrogen are entirely consumed with a brilliant light, when inflamed and exposed to the oxygen of the atmosphere. It is the residuum of the volatile portion of these substances, after driving off the gas, which makes the ammoniacal liquor so useful as a manure; all the nitrogen, with a part of the hydrogen, remaining in the liquid. In combination with chlorine, one of the elements of salt, it constitutes the muriatic, one of the strongest of the acids.

Ammonia.—The most frequent condition in which hydrogen is mentioned, in connection with vegetation, besides water, is when combined with nitrogen, in the proportion of three of the former in bulk, to one of the latter; and by weight, 17.47 of the first, to 82.53 of the last, in every 100 parts, composing the volatile alkali, ammonia, which is about $\frac{6}{10}$ the density of the atmosphere. By strong compression at a low temperature, it may be condensed to a liquid, having rather more than $\frac{3}{4}$ the specific weight of water. It is never found in a tangible shape, except in combination with acids, formimg carbonates, nitrates, sulphates, and muriates of ammonia.

Nitrogen exists in the atmosphere to the extent of about 79 per cent. The principal purpose it appears to fulfil in this connection, is in diluting the oxygen, which in its pure state, acts with too great intensity on animal life, in combustion, and all its varying combinations. So great is the attraction of undiluted oxygen for iron, that a wire, plunged into a jar of oxygen gas and ignited by a taper, will readily take fire and melt into irregular drops. This is nothing more than an illustration of the principle, exhibited in an intense degree, in the gradual rusting which takes place in the air at its ordinary temperature; or the more rapid formation of the scales under the heat of the blacksmith's forge.

All are simple oxidations of the metal, or the combination of oxygen with iron; and we see in the comparison, the immensely-accelerated effect produced by the absence of nitrogen and an augmented temperature.

Nitric acid is another compound of great importance to vegetation. It is simply nitrogen and oxygen, the identical materials which compose the atmosphere, combined in different proportions, 26.15 parts by weight of the former, and 73.85 of the latter, in every 100. This acid, in union with potash, forms nitrate of potash or saltpetre; and with soda, forms nitrate of soda. The last is found in immense beds, and lies upon and immediately under the surface of the earth, in China, India, Spain and elsewhere. From Chili it is exported in large quantities; and has been of late years, extensively used in England, as a manure.

It has been deemed relevant to our subject, to say thus much, respecting some of the most striking characteristics of those four simple principles, which make up an average of more than 98 per cent. of all living vegetables. And here, a moment's reflection irresistibly forces from us, an expression of wonder and admiration at that Wisdom and Omnipotence, which, out of such limited means, has wrought such varied and beautiful results. Every plant that exists, from the obscure sea-weed 100 fathoms below the surface of the ocean, to the lofty pines that shoot up 300 feet in mid-air; and from the clinging moss that seems almost a part of the rock on which it grows, to the expanded banyan tree of India, with its innumerably-connected trunks, overshadowing acres; every thing that is pleasant to the taste, delightful to the eye, and grateful to the smell, equally with whatever is nauseous, revolting and loathsome, are only products of the same materials, slightly differing in association and arrangement.

BARN-YARD MANURE.

The first consideration in the management of manures is, to secure them against all waste. The bulk, solubility, and peculiar tendency to fermentation, of barn-yard manure, renders it a matter of no little study, so to arrange it, as to preserve all its good qualities, and apply it, undiminished, to the soil. A part of the droppings of the cattle, are necessarily left in the pastures or about the stacks where they are fed; though it is better, for various reasons, that they should never receive their food from the stack. The manure thus left in the fields, should be beaten up and scattered with light, long-handled mallets, immediately after the grass shoots ir the spring, and again before the rains commence in the autumn, With these exceptions, and the slight waste which may occur in driving cattle to and from the pasture, all the manure should be dropped either in the stables or yards. These ought to be so arranged, that cattle may pass from one directly into the other; and the yard should, if possible, be furnished with running water. There is twice the value of manure wasted annually, on some farms, in sending the cattle abroad to water, that would be required to provide it for them in the yard for fifty years.

Keep the premises where the manure is dropped, as dry as possible; and for this purpose, the eaves may project several feet beyond the sides of the building, so as to protect the manure thrown out of the stables from the wash of rains. The barns and all the sheds should have eave troughs to carry off the water, which if saved in a sufficiently capacious cistern, would furnish a supply for the cattle. The form of the yard ought to be dishing towards the centre; and if on sandy or gravelly soil, it should be puddled or covered with clay to prevent the leakage and escape of the liquid manure. The floors of the stables may be so made, as to permit the urine to fall on a properly prepared bed of turf, placed under them for its reception, by which it would be effectually retained, till removed; or it should be led off by troughs into the yard, or what is more desirable, to a muck heap.

It is better to feed the straw and coarse fodder, which can always be advantageously done by cutting them with a straw-cutter, and mixing it with meal or roots. When it is not thus consumed, it may first be used as litter for the cattle, and as it becomes saturated with the droppings it should be thrown into the yard.

If the cattle are fed under sheds, the whole surface ought to be covered with such straw and refuse forage as can be collected; and if there is a deficiency of these, peat or any turf, which is well filled with the roots of grass, and especially the rich wash from the road side may be substituted The manure may be allowed to accumulate through the winter, unless it be more convenient to carry it on to the fields When the warm weather appproaches, a close attention to the manure is necessary. The escape of the frost permits circulation of the air through it, and the increasing heat of the sun promotes fermentation and decomposition.

Long and Short Manure.— The question has been often

mooted, as to the comparative advantages of long and short manure—*the fermented and unfermented*. This must depend on the use for which they are designed. If intended for the garden beds, for loose, light soils, or as a top dressing for meadows or any crops, or if needed to kill any noxious seeds incorporated with the heap, it should be fermented; if for hoed crops in clay or loamy soils, it should be used in as fresh or unchanged a condition as possible. Loose soils are still further loosened for a time, by long manure, and much of its volatile parts is lost before it is reduced to mold. Adhesive and compact soils, on the contrary, are improved by the coarsest manures. These tend to the separation of the earth; and all the gases which are set free in fermentation, are combined and firmly held in the soil.

Decomposition of Manures.—Three conditions are essential to produce rapid decomposition in manure; air, moisture, and a temperature above 65°; and these, except in frosty weather, are generally present in the heap. The gradual chemical changes going on in all manures, but most actively in the excrements of the horse and sheep, when they have sufficient air and moisture, induce an elevation, which keeps them always above the temperature of the surrounding air. If the manure be trodden compactly, and saturated with water, the air cannot circulate; and if its temperature be likewise kept down, it will be preserved a long time unchanged. The fermentation of manure should go forward, when thoroughly blended with all the vegetable and liquid fertilizers about the premises, and also including ashes, charcoal, gypsum and coal-dust; the last three substances combining with and retaining the ammonia as it is formed. Over all these should be placed a good coating of turf, peat, or fine mold, which will absorb any gases that escape the gypsum, and other absorbents.

Old mortar or effete lime may also be added, for the formation of nitric acid. It draws this not only from the materials in the heap, but largely also from the nitrogen of the air; it having been ascertained in the manufacture of salt-petre, (nitrate of potash,) that the amount of nitrogen in the salt, is greatly increased above that in the manure used. The absorption of nitre by lime, in a course of years, is very large, as is shown by the practice of the Chinese farmers, who to secure it for manure, will gratuitously remove the old plaster on walls and replace it with new.

If required to hasten decay and especially, if there be in

tractable vegetables, as broom and other corn-stalks, or such
as have seeds that ought to be destroyed, they may be well
moistened and thrown together in layers, three or four inches
thick; and on each may be strewn a liberal coating of fresh,
unslacked lime, reduced to powder. This promotes decomposition, and when it is far enough advanced, the whole may
be sparingly added to the general mass, as the lime will by
that time have become mild. When remote from the cattle
yard, these coarse materials may be at once burned, and the
ashes added to the soil; or they may be buried in furrows,
where the ground will not be disturbed till they are entirely
rotted.

When thoroughly decomposed, the manure heap will have
lost half its original weight, most of which has escaped as
water and carbonic acid. It may then be carted on to the
ground, and at once incorporated with it; or if intended for
a top dressing, it should be scattered over it, immediately
before or during wet weather For the protection of the
manure, it would be well to cover it with a roof, and convey
off all the water from the eaves. This will prevent any
waste of the soluble portions and promote the escape of moisture, by the free circulation of air, which to the extent of this
evaporation, will lessen the labor of hauling.

Tanks for holding Liquid Manure have long been in
use. They should be convenient to the stalls and yards, and
with tight drains, convey into them every particle of the
urine and drainage from the manure. In compact clay, they
may be made by simply excavating the earth, and the sides
can be kept from falling in, by a rough wall, or by planks supported in an upright position, by a frame-work of joice. But
in all cases, the cisterns should be *closely covered*, to prevent
the escape of the ammonia, which is developed while fermenting. In porous soils, it is necessary to construct them
with stone or brick, laid in water-lime or cement.

When partially filled, fermentation will soon take place in
the tank, and especially in warm weather; gypsum or charcoal should then be thrown in to absorb the ammonia. A
few days after decomposition commences, it should be pumped
into casks and carried upon the land. If intended for watering plants, it must be diluted sufficiently to prevent injury
to them. The quantity of water required, will depend on
the strength of the liquid, and the time it is applied; much
less water being necessary to dilute it in a wet, than
in a dry time. By fermenting in the open air and undiluted,

it has been found, that in six weeks, cow's urine will lose nearly one-half of its solid matter or salts, and $\frac{6}{7}$ of its ammonia; while that which had been mixed with an equal quantity of water, lost only $\frac{1}{18}$ of the former, and $\frac{1}{5}$ of the latter. The stables and troughs leading to the tank, should be frequently washed down and sprinkled with gypsum. This last will absorb much of the ammonia, which would otherwise escape. Some loss of the volatile matter must be expected, and the sooner it is used after proper fermentation or *ripeness*, as it is termed, the greater will be the economy.

Liquid Manure applied to the Muck Heap.—As a general rule, it is more economical and a great saving of labor, to keep the urine above ground, and mix it at once with the manure; but in this case vegetable or earthy absorbents must be adequately supplied; and in addition, the heap ought frequently to be sprinkled with gypsum or charcoal. Rich turf, the wash of the road-side, tan bark or saw-dust, and all refuse vegetables may be used for this purpose, being so placed that the liquid can run on to them, or be deposited where it can be poured over the heap. The same protection, of a rough, open shed, should be given to this, as to the other heaps, to facilitate evaporation and prevent drenching from rains. When fully saturated with the urinary salts, and all is properly decomposed, it may be carried out for use, or closely covered with earth till wanted. The decomposition is in a great measure arrested, by covering with compact earth, thoroughly trodden together; this prevents the access of air, which is essential to its progress.

A simple yet economical mode of saving the liquid manure, is sometimes adopted in Scotland, and is thus detailed:

" We divide a shed into two compartments; one of which we make water-tight, by puddling the side walls with clay to the height of two feet, and separated from the other compartment by a low, water-tight wall or boarding. This is the fermenting tank, which is filled half or three-fourths full of pulverized burnt peat, and the liquid manure from the stables and pig-styes, directed into it. This is mixed up with the pulverized peat, and allowed to remain three or four weeks, till the decomposition seems about completed, being occasionally stirred after the composition has become about the consistency of gruel. The whole is then ladled with a pole and bucket, over the low partition into the second floor, which is also three parts filled with the carbon ized peat; and as the second floor is meant merely as a filter we have it lower on one side than the other, by which means in the course of a day or two, the carbonized peat is left comparatively dry. The water having passed off at the lower side, the first or fermenting floor is again filled as before, and the contents of the second floor, if considered saturated enough, are then shovelled up into a corner, and allowed to dry till used, which may be either immediately, or at the end of twenty years, as scarcely anything will effect it, if not exposed to the continued washing of pure water, or exposed to the influence of the roots of growing plants. By being thinly spread on a granary floor, it soon becomes perfectly dry, and suited to pass through drill machines.

The mixing of the carbonized peat with the liquid manure, on the first or fermenting floor, is for laying hold of the gaseous matters, as they escape during the fermentation; perhaps other substances may secure these more effectually, but none so cheaply. By this plan, a great many desiderata are at once obtained. You get free of over 900 parts out of every 1,000 of the weight and bulk of manure, by the expulsion of the water; while at the same time, all the fertilizing properties contained in it, are combined with light, cleanly, and portable materials, and possessed of the peculiar property of holding together the most volatile substances, till gradually called forth by the exigencies of the growing plants. Lastly, you get free of the tank, hogshead, and watering cart, with all its appendages, and are no more bothered with overflowing tank, or overfermented liquid, with weather unsuited to its application. You have merely to shovel past the saturated charcoal, and shovel in a little fresh, and the process goes on again, while the prepared peat lies ready for all crops, all seasons and all times."

Value of Liquid Manures.—The urine voided from a single cow, is considered worth $10 per annum, in Flanders, where agricultural practice has reached a high state of advancement. It furnishes 900 lbs. of solid matter, and at the price of $50 per ton, for which guano is frequently sold, the urine of a cow for one year is worth $20. And yet economical farmers will continue to waste urine and buy guano! "The urine of a cow for a year will manure 1¼ acres of land, and is more valuable than its dung, in the ratio by bulk, of seven to six; and in real value as two to one."— (*Dana*) How important, then, that every particle of it be carefully husbanded for the crops.

The average urine of the cow, as analyzed by Sprengel, contains 92.6 per cent. of water ; that of the horse, 94 ; the sheep, 96 ; the hog, 92.6 ; and the human, 93.3. The re-mainder is composed of salts and rich food for vegetables; but the human is far richer in these than any other. The quantity and value of urine, varies much in different speci-mens from the same or similar subjects, and depends on the food and liquid taken into the stomach, the loss by perspira-tion and other circumstances.

SOLID ANIMAL MANURES AND THEIR TREATMENT.

Of these, *Horse dung* is the most valuable and the easiest to decompose. If in heaps, fermentation will sometimes com-mence in 24 hours; and even in mid-winter, if a large pile of fresh manure be accumulated, it will proceed with great rapidity. If this is not arrested, a few weeks, under favorable circum-stances, are sufficient to reduce it to a small part of its origi-nal weight and value. Boussingault, one of the most care-ful observers of nature, as well as an accurate, experimental chemist, states the nitrogen in fresh dried—horse dung to be 2.7 per cent. of its whole weight. The same manure laid in a thick stratum and permitted to undergo thorough de-composition, loses $\frac{9}{10}$ of its entire weight, and the remain-ing tenth when dried, gives only one per cent. of nitrogen. Such are the losses which follow the neglect of inconsider-ate farmers. Peculiar care should therefore be taken, to arrest this action at the precise point desired. Salt scattered through the heap, will materially lessen the activity of de-composition. It is better, however, to add turf as it accu-mulates, in addition to the salt, if it is to remain long before being composted or carried on the land.

The manure of Sheep is strong and very active, and next to that of the horse, is the most subject to heat and decom-pose. *The manure of Cattle and Swine* being of a colder nature, may be thrown in with that of the horse and sheep in alternate layers, or it may remain in heaps by itself if more convenient.

If fresh manure be intermixed with straw and other ab-sorbents, (sea-weed, peat, turf, tan-bark and the like,) and over this a thick covering of earth or peat be placed, this ex-ternal coating will combine with any volatile matters, which fermentation developes in the lower part of the mass, and preserve most of it from waste. Frequent turning of the manures, is a practice attended with no benefit, but with the certainty of the escape of much of its valuable properties.

Many farmers assign a distinct or peculiar merit to the different manures. Much of this opinion is fanciful; for there is frequently more difference in the comparative value of that from the same species, and even the same individual, at different times and under different circumstances, than from those of different species.

The diversity in manures may arise from several causes. The more thoroughly the food is digested and its nutritive qualities extracted, the less is the value of the manure. Thus, on the same quantity and quality of food, a growing animal, or a cow in calf or giving milk, yields a poorer quality of feces and urine, than such as are not increasing in weight ; and if the animal be actually losing condition, the richness of the manure is very much increased.

The quality of food adds materially to this difference, the richest giving by far the most valuable manure. Those animals which are kept on a scanty supply of straw or refuse hay, yield manure little better than good turf, and far inferior to the droppings of such as are highly fed. The imperfect mastication and digestion of the horse and mule, in comparison with the ruminating animals, the ox and sheep, their generally better quality of food, and the fact, that for the greater part of their lives they are not adding to their carcass, is the cause of the increased value of their manure. Their solid feces are also much richer than those of the cow, as they void less urine and this is of an indifferent character. In a long series of careful experiments, made at Dresden and Berlin by order of the Saxon and Prussian governments, it was ascertained, that unmanured soil which would yield three for one sown, when dressed with cow dung, would give seven ; with horse dung, ten; and with human, fourteen.

For the purpose of showing the proportions of the various elements which compose the farmer's manure heap, an analysis from Mr. Richardson is subjoined, of some taken from the farm yard in the condition usually applied to the field.

	Fresh		Dried at 212°.
Water	64.96	Carbon .	. 37.40
Organic matter	24.71	Hydrogen .	. 5.27
Inorganic salts	10.33	Oxygen .	. 25.52
		Nitrogen .	. 1.76
	100.00	Ashes	. 30.05
			100.00

INORGANIC MATTERS.

Portion soluble in muriatic acid.		Soluble in water	
Silica	. 27.01	Potash .	. 3 22
Phosphate of lime .	7.11	Soda .	. 2.73
Phosphate of magnesia	2.26	Lime .	. 0.34
Phosphate of iron .	4.68	Magnesia	. 0.26
Carbonate of lime .	9.34	Sulphuric acid	. 3.27
Carbonate of magnesia	1.63	Chlorine .	. 3.15
Sand . .	30.99	Silica .	. 0.04
Carbon . .	83		___
Alkali, and loss .	3.14		13.01
	___		86.99
	86.99		___
			100.00

The following is from other specimens of fresh **farm** yard manures, analyzed by Messrs. Allen and Greenhill.

				Farmyard Manure from Kent.	Farmyard Manure from Surrey
Per-Centage of Ash - - -				9.2	9.6
Silica - - . - . -				70.79	71.32
Potash - - - - . -				3.32	5.14
Soda - - - - - -				0.92	1.68
Lime . - - - - -				6.90	12.32
Magnesia - - - - -				0.56	0.82
Common salt - - - -				1.43	1.22
Phosphate of iron - - -				2.04	2.03
Phosphate of alumina - - - -				1.53	2.54
Sulphuric acid - - - - -				1.89	1.57
Phosphoric acid - - - -				1.58	1.27
Manganese - - - - -				a trace	___
				99.76	99.91

By knowing the composition of the added manures **and** the subtracted crops, the farmer can keep an intelligent account of debt and credit with his fields. If he could make an exact estimate of the portion of the soil that might become soluble in the course of the growing season, (available for the present crop), and carry into this account also, the sum of the elements exhausted by drainage and evaporation, as well as those added from the atmosphere, rains and dews, and appropriated to vegetation or permanently fixed in the soil, he would then be able, at all times to know, precisely what additional ingredients (*special manures*) would be necessary, and in what proportions, to secure the largest amount of any required crop.

POUDRETTE AND URATE.

Poudrette is the name given to the human feces after being mixed with charcoal dust or charred peat. By these it is disinfected of its effluvia, and when dried, it becomes a convenient article for use, and even for remote transportation. The odor is sometimes expelled by adding quick lime, but this removes with it much of the ammonia, and on this account should always be avoided.

Urate as well as poudrette, has become an article of commerce. It is manufactured in large cities by collecting the urine, and mixing with it 1-6th or 1-7th of its weight of ground gypsum, and allowing it to stand several days. The urine combines with a portion of the ammonia, after which it is dried and the liquid is thrown away. Only a part of the value is secured by this operation. It is sometimes prepared by the use of sulphuric acid, which is gradually added to urine and forms sulphate of ammonia, which is afterwards dried. This secures a greater amount of the valuable properties of the urine ; but even this is not without waste.

Night Soil.—*From the analysis* of Berzelius, the excrements of a healthy man, yielded water, 733 ; albumen, nine ; bile, nine ; mucilage, fat and the animal matters, 167 ; saline matters, twelve ; and undecomposed food, 70—in 1,000 parts. When freed from water, 1,000 parts left, of ash, 132 ; and this yielded, carbonate of soda, eight ; sulphate of soda, with a little sulphate of potash, and phosphate of soda, eight ; phosphate of lime and magnesia, and a trace of gypsum, 100 ; silica, sixteen.

Human urine, according to the same authority, gives in every 1,000 parts, of water, 933 ; urea, 30.1; uric acid, 1.0 ; free acetic acid, lactate of ammonia, and inseparable animal matter, 17.1 ; mucus of the bladder, 0.3 ; sulphate of potash, 3.7 ; sulphate of soda, 3.2 ; phosphate of soda, 2.9 ; phosphate of ammonia, 1.6 ; common salt, 4.5 ; sal-ammoniac 1.5 ; phosphates of lime and magnesia, with a trace of silica and of fluoride of calcium, 1.1.

Urea is a solid product of urine, and according to Prout, gives of carbon, 19.99 ; oxygen, 26.63 ; hydrogen, 6.65 ; nitrogen, 46.65—in 100 parts. The analysis of Wœhler and Liebig differs immaterially from this. Such are the materials, abounding in every ingredient that can minister to the production of plants, which are suffered to waste in the air, and taint its purity and healthfulness ; or they are buried deep in the earth beyond the reach of any useful application ; and even in this position, (frequently in villages, and always in cities), they pollute the waters with their disgusting and poisonous exudations. The water from one of the wells in constant use in Boston, examined by Dr. Jackson, gave an appreciable per centage of night soil

Treatment of Night Soil.—No perfect mode has yet been devised of managing night soil. For compactness and facility of removal, we suggest, that in cities, metallic boxes of sufficient capacity be placed in the privies, so arranged as to be easily taken out in the rear, for the purpose of empty ing their contents. To prevent corrosion, they may be made of composite or galvanized metal. In the country where it can be at once applied, tight wooden boxes may be used with hooks on the outer side, to which a team may be attached for drawing it out wherever required. The boxes should have a layer of charcoal dust, charred peat or gypsum at the bottom, and others successively as they become filled. These materials are cheap, compact, and readily combine with the volatile gases. Sulphuric acid is more efficient than either, but more expensive. Quick-lime will neutralize the odor, but it expels the enriching qualities; and if it be intended to use the night soil, lime should never be mixed with it. Both the charcoal and peat condense and retain the gases in their pores, and the sulphuric acid of the gypsum leaves the lime, and like the free acid, combines with the ammonia, forming sulphate of ammonia, an inodorous and powerful fertilizer. Raw peat, turf, dry tan-bark, saw dust and ashes are all good ; but as more bulk is needed to effect the object, their use is attended with greater inconvenience. From its great tendency to decompose, night soil should be immediately covered with earth, when exposed to the air. It is always saved by the Flemings and Chinese ; the former generally using it liquid, and the latter, either as a liquid, or mixed with clay and dried like brick.

The sole use of this manure, guano, ashes, charcoal, lime, gypsum and other salts, effectually prevents the propagation of all weeds. Its value, like all others, depends much on the food from which it is derived ; it being richest when large quantities of meat and other nutritive food is consumed. The difference in the products from the best hotels and poorly supplied work-houses, though not in proportion to the first cost of the food consumed, yet bears no inconsiderable ratio to it.

THE EXCREMENTS OF FOWLS.

These contain both the feces and urine combined, and are next to night soil in value. They should be kept dry, or what is better, mixed at once with the soil, or with a compost where their volatile matters will be retained. They are very soluble, and when exposed to moisture, are liable to waste. Since these contain the essential elements of guano, the economy of saving them must be apparent, to those who buy the imported fertilizers at so large a cost.

FLESH, BLOOD, &c.

When decomposed, these substances afford all the materials of manure in its most condensed form. Whenever procurable, they should be mixed with eight or ten times their weight of dry peat, turf, tan-bark or rich garden mold. A dead cow or horse thus buried in a bed of peat, will yield 12 or 15 loads of the richest manure. Butchers' offal when thus treated, will yield ten times its weight of more valuable manure than any from the cattle yards.

HAIR, BRISTLES, HORNS, HOOFS, PELTS, THE FLOCKS AND WASTE OF WOOLEN MANUFACTORIES AND TANNERIES.

These are all rich in every organic material required by plants; and when mingled with the soil, they gradually yield them, and afford a permanent and luxuriant growth to every cultivated crop. Most animal substances contain from 15 to 18 per cent. of nitrogen; and when it is considered, that this is a greater amount than is afforded by an equal quantity of saltpetre, and about two-thirds of that contained in nitric acid, (one of the most condensed and powerful manures), the recklessness and waste is apparent, of throwing dead animals and similar manures by the road-side, and allowing them to decay above ground; thus robbing the soil of its just dues, and afflicting the nostrils of the community with what if rightly appropriated, might minister to the necessaries and even to the luxuries of mankind.

FISH.

Fish are extensively used in this and other countries for manure. The moss-banker, alewives and other fish frequent the Atlantic coast in countless numbers in the spring and summer, and are there caught in seines, and sold to the farmers by the wagon-load. They are sometimes plowed into the soil with a spring crop; or they are more frequently used for growing corn, for which purpose, one or two fish are placed in each hill and buried with the seed; or they are turned under near the young corn, at the first running of the plow in cultivating. This was the system adopted by the Aborigines of our country, in raising their maize on exhausted lands, long before their occupancy or even discovery by the whites. There is waste in this practice, as the soils used for corn are generally light and sandy ; and the slight, silicious covering imperfectly combines with the putrefying fish, and much of their gases thereby eludes the plant, to the excessive annoyance of the olfactories of the residents, for miles around.

The proper method of using them, is by composting with dry peat, in alternate layers of about three inches in thickness of fish to nine of peat, and over the whole, a coating of two or three feet of peat is placed. A few weeks of warm weather suffice to decompose the fish, which unite with the peat, no perceptible effluvia escaping from the heap, so effectual is its absorption. A strong acid smell is, however, noticeable, originating in the escape of the acidifying or antiseptic principle contained in the peat, which has kept it for ages in a state of preservation, and whose expulsion is the signal for breaking up its own structure. It now passes rapidly into decay, and is soon lost in a mass of undistinguishable, vegetable mold, the fruitful bed of new and varied vegetable forms. This compost may remain without injury or waste for years. Two or three weeks before using, it should be overhauled and intimately mixed, when another fermentation commences with an elevation of temperature. When this ceases, it may be applied to the land. This compost will be found adapted to nearly all soils and crops.

COTTON SEED.

This is yielded at the rate of 200 to 400 lbs. per acre, and is a valuable manure. It would doubtless be more profitable if first made to yield the oil which it contains, to the amount of about 20 per cent. and use the residuum as a manure ; or to feed it, when properly prepared, to the stock, and use their manure for the fields. Where this is not done, however, the seed ought carefully to be saved and applied to the land, at the rate of 60 to 80, or even 100 bushels per acre.

It may be scattered broadcast, and plowed in during the winter, where it will rot before spring ; or it may be thrown into heaps and allowed to heat, and when vitality is destroyed, it may be plowed or drilled in, or thrown upon the corn hills and buried with the hoe or plow.

SEA WEED

Is a powerful aid to the farmer, when within convenient distances. It is thrown upon the sea-coast by the waves in large winrows; or it is carefully raked up from the rocks or bottom of the bays, either by farmers or those who make it a business to procure and sell it. It may be used as bedding for cattle or litter for the barnyard, or added directly to the compost heap. Where the distance for carrying it would prevent its use, it may be burned, and the ashes removed to the land. It has much more saline matter than vegetables which grow on land, and yields a more valuable manure.

PEAT.

This substance is seldom found in this country, in the purity that characterizes it in many parts of northern Europe. There, its nearly pure carbonaceous quality admits of its extensive use as fuel. In the United States, it is generally mixed with the wash from the adjacent elevations, which renders it more easily susceptible of profitable cultivation in its native bed, and scarcely less valuable as a fertilizer when applied to other lands. In six different specimens from Northampton, and four from other localities in Massachusetts, Dr. Dana found an average of 29.41 soluble, and 55.03 insoluble vegetable matter; and 15.55 of salts and silicates, in every 100 parts. His researches have led him to recommend the mixture of 30 lbs. potash, or 20 lbs. soda ash; or what is more economical and equally efficacious, eight bushels of unleached wood ashes with one cord of peat as it is dug from its bed; or if leached ashes be used, they should be mixed in the proportion of one to three of peat. This he considers fully equivalent to pure cow dung in value. He also estimates the salts and organic matter of four cords of peat, as equal to the manure of a cow for one year. The opinion of Mr. Phinney, of Lexington, Mass., founded on close observation and long practice, is, that one part of green cattle dung, composted with twice its bulk of peat, will make the whole equal in value to the unmixed dung.

Peat in its natural condition, contains from 70 to over 90 per cent. of water. It should be dug from its bed in the fall or winter, for the purpose of draining and exposing it to the action of the atmosphere, when it will be found to have lost about two-thirds of its bulk. In this state, it still holds about 65 per cent. of water. It may then be carried into the cattle yards, and used for making composts in any way desired.

RICH TURF.

Much of this is full of the roots of grasses and decayed vegetables, and is a valuable absorbent of every species of animal or other manures. Whenever it can be procured oy the road-side or other waste places, it should be used for this purpose. It is frequently filled with the seeds or roots of weeds, which ought to be killed preparatory to using as manure. A mutually beneficial effect is produced, by mixing turf with lime, by which the turf is speedily rotted, and the obnoxious weeds killed ; and the lime is thus becoming equally fitted to act beneficially when applied to the soil, as if already incorporated with it. Some weeks after mixing together, the heap will be in a fit condition to receive every description of manure.

SWAMP MUCK OR POND MUD.

Under certain conditions, this is a more valuable addition to the muck heap, or more properly, a foundation for it, than either of the preceding. Especially is this the case, when there is no outlet for the water and sediment ; and the mud, besides containing a large proportion of salts, the result of ages of evaporation, is the receptacle of the remains of myriads of minute shell fish, animalculæ, infusoriæ, the spawn and exuviæ of frogs and other occupants. Ducks and various aquatic birds fill themselves to repletion, when ranging through a pond thus daintily supplied; the contents of which are even much more adapted to the promotion of vegetable than animal life. Such reservoirs of vegetable nutrition, are mines of wealth to the farmer, if judiciously applied ; nor can he justify meagre returns from his fields, while this remedy is within his reach.

MANURING WITH GREEN CROPS.

This system has within a few years, been extensively adopted in some of the older-settled portions of the United States. The comparative cheapness of land and its products, the high price of labor, and the consequent expense of making artificial manures, renders this at present, the most economical plan which can be pursued. The design in this practice is primarily, fertilization ; and connected with it, is the clearing of the ground from noxious weeds, as in fallows, by plowing in the vegetation before the seed is ripened ; and finally, the object is to loosen the soil and place it in the mellowest condition for the crops which are to succeed. Its re-

4

sults have been entirely successful, when steadily pursued, and with a due consideration of the objects sought and the means by which they are to be accomplished.   Lands in many of our eastern States, which have been worn out by improvident cultivation, and unsalaeble at $10 an acre, have by this system, while steadily remunerating their proprietors by their returning crops, for all the outlay of labor and expense, been brought up in value to $50 per acre.

The full benefit of green crops as manures seems only to be realized where there is sufficient lime in the soil.   Calcareous soils, or such as have a large proportion of lime, however they may have become exhausted, when put under a thorough course of treatment, in which green crops at proper intervals are returned to them, are soon restored to fertility ; and when lime does not exist in the soil, the application of it in the proper manner and quantity, will produce the same effect.   Gypsum and ashes are the best substitutes, when lime or marl is difficult to be procured.

This system of improvement, varies with almost every individual who practices it, according to the quality of his land, the kind of crops to be raised, the facility of procuring manures, the luxuriance of particular crops, and other considerations.   We shall state merely, the general principles in this, as in most other subjects, and leave to the farmer's judgment, to apply them according to his circumstances.   It is always better to commence this system, while the land is in good condition, as a luxuriant growth of vegetation is as profitable for turning in, as for cropping.   Buckwheat, oats, rye, and some of the grasses, have been used for this purpose in this country ; and spurry, the white lupine, the vetch and rape in Europe ; but for the northern portion of the Union, nothing has been hitherto tried, which is so well fitted for the object, as red clover.

CLOVER FOR GREEN MANURES.

This is suited to all soils, that will grow anything profitably, from sand, if possessing an adequate amount of fertility, to the heaviest clay, if drained of its superfluous water. The seed is not expensive, its growth certain and rapid, and the expense of its cultivation trifling ; while the return, on a kindly soil, and with proper treatment, is large.   Added to this, and very much increasing its merits, is the abundance of its long tap roots, which penetrate the ground to a great depth, and break up the stiff soils, in a manner peculiarly beneficial to succeeding crops. The material yielded by the roots and stubble, is of itself equal to a good dressing of manure. It has the further advantage, of giving two or more years of growth from one sowing, and of maintaining itself in the ground thereafter, by self seeding, when not too closely cropped; and it is equally suited to profitable pasturage and winter forage.

If the first season's growth be luxuriant, after the removal of the grain upon which it was sown, clover may be pastured in the autumn, or suffered to fall and waste on the ground, the first being the most economical. The following year, the early crops may be taken off for hay, and the second, after partially ripening its seeds, may be plowed in; and thus it carries with it, a full crop of seed for future growth. It is usual when wheat is cultivated, to turn in the clover when in full flower in July, and allow the ground to remain undisturbed till the proper time for sowing the grain; when it may be cross-plowed if necessary, or the wheat may be sown directly on the ground and harrowed in. This system gives alternate crops of grain and clover, and with the use of such saline manures, as may be necessary to replace those abstracted from the soil; it will sustain the greatest fertility. With a slight dressing of these, when the land is in good condition, the first crop of clover may be taken off, and yet allow a sufficient growth for turning in.

It is customary, however, to adopt a three or four years course of cropping, in which grain, roots, corn, &c. alternate with clover and barn-yard manures; and this we think the most judicious practice, when the land is within convenient distance of the manure. If the fields are remote, a still longer course would be preferable, where stock and particularly sheep are kept; as they might be allowed to pasture the field during a much greater time. Sheep would remove only so much of the forage as remains in their carcass; while milch cows and working animals would, of course, carry off a greater amount, the first in the milk, and the last in their manure, dropped while out of the field.

THE COW PEA,

Like the pole bean, of which it seems to be a kindred genus, grows with a long vine and abundance of leaves. It is deemed the best of the fertilizers for the South. It will there mature in the same field with the corn, after that has ripened; or it will grow two crops in one season, from two successive plantings. This is also a valuable fodder for cattle and sheep, and the ripe peas are a profitable crop. Like a luxuriant growth of clover, it requires the roller to prepare it properly for the plow, when turned under previous to the decay of the vines. The cow pea is an economical fertilizer, in consequence of its broad, succulent, bean-like leaves, drawing nitrogen and carbonic acid largely from the air, but its slight fusiform roots, do not effect that mechanical division so beneficial to many adhesive soils, which is produced by the long tap roots of the clover.

Spurry is extensively used as a fertilizer in the north of Europe, (Flanders, Germany and Denmark), and as forage for cattle, both in its green and dry state. It is admirably adapted to the lightest sands, where it is said to grow with more luxuriance and profit, than any other of the cultivated plants. It may be sown in the fall, after grain or early roots, and plowed under the following spring. Three crops may be grown on the same land in one season. Van Voght says, by alternating these crops with rye, it will reclaim the worst sands, and yield nearly the same benefits, if pastured off by cattle ; while it adds materially to the advantages of other manures applied at the same time. It grows spontaneously in many of our fields, as a weed ; and its cultivation on our lightest sands, which are too poor for clover, might be attended with the best effects. Like the cow pea, however, it is deficient in the deep, tap roots, which give much of their efficiency to the clover and white lupine.

WHITE LUPINE (Lupinus albus.)

This plant has not, to my knowledge, been introduced as a field crop in this country ; but from the great success which has attended its cultivation in Europe, it is a proper subject of consideration, whether it might not be advantageously introduced among us as a fertilizer. It grows freely in all except calcareous soils, and is best suited to such as have a subsoil charged with iron. It is hardy, not liable to injury from insects, grows rapidly and with an abundance of stems, leaves and roots. The latter protect the plants from drought, by penetrating through the subsoil for a depth of more than two feet, which they break up and prepare, in the most efficient manner, for succeeding crops

THE ADVANTAGES OF GREEN MANURES

Consist principally in the addition of vegetable matter,

which they furnish to the soil. The presence of this, aids in the liberation of those mineral ingredients, which are there locked up, and which on being set free, act with so much advantage to the crop. The roots also, exert a power in effecting this decomposition, beyond any other known agents, either of nature or ai:. Their minute fibres are brought into contact with the elements of the soil, and they act upon them with a force peculiar to themselves alone. Their agency is far more efficacious for this purpose than the in tensest heat or strongest acids, persuading the elements to give up for their own use, what is essential to their maturity and perfection. By substituting a crop for a naked fallow, we have all the fibres of the roots throughout the field, aiding the decomposition which is slowly going forward in every soil.

Clover and most broad-leaved plants, draw largely for their sustenance from the air, especially when aided by the application of gypsum. By its long tap roots, clover also draws much from the subsoil; as all plants appropriate such saline substances as are necessary to their maturity, and which are brought to their roots in a state of solution, by the up-welling moisture from beneath. This last is frequently a great source of improvement to the soil. The amount of carbon drawn from the air in the state of carbonic acid, and of ammonia and nitric acid, under favorable circumstances of soil and crop, is very great; and when buried beneath the surface, all are saved and yield their fertility to the land; while such vegetation as decays on the surface, loses much of its value by evaporation and drainage. In the green state, fermentation is rapid, and by resolving the matter of plants into their elements, it fits the ground at once for a succeeding crop.

Additional manures cannot be more particularly specified here. It is sufficient to add, that every portion of vegetable or animal substances, and many which are purely mineral, may be used on the fields with the utmost advantage to the farmer. Intelligent observation, experience, and that knowledge which he will acquire from the best modern agricultural writings of the present day, will enable him to adapt them in the most judicious manner, to his soil and crops.

THE FALLOW SYSTEM.

As a means of enriching lands, this was formerly much practised, but it is now entirely discarded by intelligent farmers. It consists in plowing up the land and exposing it naked to the elements, whenever the exhaustion by tillage requires it. This practice is founded on the principle, that plants gradually exhaust the soil of such soluble food, potash, soda, and other materials, as are necessary to their support; and unless they are again given to it in manures, in a form suited to their immediate appropriation by plants, time is requisite for dissolving them in the soil, so as to enable them again to support vegetation profitably. Besides the loss resulting from the frequent idleness of the land, naked fallows have this further disadvantage, and especially in light and loose soils; they are exposed to the full action of the sun and rains, and by evaporation and drainage, are exhausted of much of their soluble, vegetable food.

This system, bad as it is, may yet be absolutely necessary, where grain alone is raised and no manure is applied. But it is always avoidable, by substituting fallow crops, as they are termed, potatoes, swede turneps, and other well-hoed crops, with manure; or clover, or other green crops, as above detailed; by which the land is cleared of weeds and sufficiently enriched for succeeding cultivation. If they have been kept in good condition by top-dressing, meadows are equally fitted for the different species of grains or other crops, as if the land had been fallowed; and pastures answer the same purpose, without the aid of other manures than such as have incidentally accumulated upon them.

CHAPTER IV.

IRRIGATION AND DRAINING.

IRRIGATION may properly enough be classed under the head of manures; as the materials which it provides are not only food for plants, but they aid also in procuring it from other sources. Water is of indispensable necessity to vegetable life; and the great quantity of it demanded for this purpose, is in most climates, amply provided by nature in the stores of rain and dew which moisten the earth, and especially during the early growth of vegetation, when it is most required. In countries where rain seldom or never falls, as in parts of South America, Egypt, and elsewhere, the radiation of heat from the surface, is so rapid under their clear skies, that excessive deposites of dew, generally supply the plants with all the moisture which they need. The same effect takes place in our transparent, summer atmosphere, throughout most of the United States; and it is to the presence of copious dews, on our rich, well cultivated fields, that much of the luxuriance and success is due, which has ever attended enlightened and judicious American husbandry.

Besides the moisture that abounds in the atmosphere, (but which is not always available in rains, and dews to the extent desired for the wants of vegetation), and that which imperceptibly ascends from remote depths in the earth, and contributes to the support of plants; it is a practice coeval with the earliest history of agriculture, to bring artificial waters upon the cultivated fields and make them tributary to the support of the crops. In many countries this system is indispensable to secure their maturity; for although dews accomplish the object in a measure, they do not supply it in the quantity required to sustain a vigorous growth. We find, in looking to the practice of Egypt and the Barbary States in Africa; of Syria, Babylon, and other parts of Asia; Italy, Spain and elsewhere in Europe, in each of which husbandry early attained a high rank, that irrigation

was extensively introduced. Damascus is one of the most
ancient cities on record, (for it is mentioned in Genesis as
existing nearly 4,000 years ago); and notwithstanding its
numerous successive masters, and its frequent plunder and
devastation, it is still a flourishing city, though in the midst
of deserts. This is no doubt owing to the waters derived
from the " Abana and Pharpar, rivers of Damascus," which
are conducted above the city, where they gush from the
fountains, and thence overspread the gardens and water all
the adjacent plain. Had it not been for irrigation, Damascus would doubtless, ages ago, have followed Palmyra, the
Tadmor of the wilderness, into utter abandonment and ruin.
On no other principle than a systematic and extensive practice of irrigation, can we account for the once populous
condition of Judea, Idumea, and other vast regions in the
East; many of which, to the eye of the modern traveller,
present nothing but the idea of irreclaimable sterility and
desolation.   The possession of the " upper and nether
springs," was as necessary to the occupant, as possession
of the soil.

In those countries where the drought is excessive, and
rains are seldom to be depended upon, water is led on to the
fields containing all the cultivated crops, and is made subser-
vient to the growth of each. But in the United States, and
in the middle and northern part of Europe, where the vegetation ordinarily attains a satisfactory size without its aid,
irrigation is confined almost exclusively to grass or meadow
lands.

All waters are suitable for irrigation, excepting those
containing an excess of some mineral substances, deleterious
to vegetable life. Such are the drainage from peat swamps,
from saline and mineral springs, and from ore beds of various
kinds; but those are most frequent, in which iron is held in
solution. Of the spring or ordinary river waters, those are
the best which are denominated *hard*, and which owe this
quality to the presence of sulphate or carbonate of lime or
magnesia.   Such waters as are charged with fertilizing substances, that have been washed out of soils by recent floods,
are admirably suited to irrigation.   Dr. Dana, estimates the
quantity of salts in solution, and geine or humus (vegetable
matters), which were borne sea-ward past Lowell, on the
Merrimac River in 1838, (a season of unusual freshets), as
reaching the enormous amount of 840,000 tons; enough to
have given a good dressing to 100,000 acres of land. Turbid

waters that have flowed out of the sewers of cities, or past slaughter-houses and certain manufactories, and received the rich contributions of vegetable food thereby afforded, are the most beneficial. Meadows thus irrigated, in the neighborhood of Edinburgh, have rented by the acre, at the large sum of $250 per annum; a price predicated not only on the enormous amount of grass yielded, but on the high prices at which it was retailed in that city. But when none of these can be procured, pure spring water, apparently destitute of any soluble matters, may be advantageously used.

Additional effects of Irrigation. Besides its drainage ot fertile matters from remote distances, which are deposited on the fields overflown, water freely absorbs the gases, (carbonic acid, oxygen and nitrogen, &c.,) in proportions altogether different from those existing in the air, and brings them to the roots, by which they are greedily appropriated; and in its onward, agitated progress over the field, it again absorbs them from the air, again to be given up when demanded by the roots. When the water is permitted to remain stagnant on the surface, this good effect ceases; and so far from its promoting the growth of the useful and cultivated grasses, they speedily perish, and a race of sour and worthless aquatic plants spring up to supply their place.

Another and important office that water fulfils in ministering to the growth of vegetation, is in disposing the soil to those changes, which are essential to its greatest fertility. Gypsum requires 460, and lime 778 times its bulk of water at 60° Far. to dissolve them. Others among the mineral constituents of plants, also require the presence of large quantities of water, to fit them for acting on the soil, and to adapt them for vegetable assimilation.

TIME FOR APPLYING WATER TO MEADOWS.

In those regions where the winters are not severe, water may be kept on the fields during the entire season of frosts. This prevents their access to the ground, and on the approach of warm weather, the grasses at once start into life, and yield an early and abundant growth. But in general, this system cannot be successfully practiced. The water may be admitted at proper intervals, freely during the spring and early part of the summer, when vegetation is either just commencing or going forward rapidly. It is sufficient to flood the surface thoroughly, and then shut off the water
4*

for a time. In very dry weather, this may be done with advantage every night. Continued watering under a bright sun, is an unnatural condition with upland grasses, and can never be long persisted in without proving fatal to them. Neither should the water be applied after the grasses have commenced ripening. Nature is the proper guide in this, as in mos of the operations of the farmer; and it will be seen, how careful she is, in ordinary seasons, to provide an affluence of rains for the commencement of vegetation, while she as carefully withholds them when it approaches maturity. Immediately after the grass is cut, the water may be again let on as occasion requires, till the approach of cold weather. Pastures may be irrigated from time to time, as the weather may demand, throughout the entire season.

THE MANNER OF IRRIGATING.

This must depend on the situation of the surface and the supply of water. Sometimes, reservoirs are made for its reception from rains or inundations; and at others, they are collected at vast expense, from springs found by deep excavations, and led out by extensive subterraneous ditching. The usual source of supply, however, is from streams or rivulets, or copious springs, which discharge their water on elevated ground. The former are dammed up, to turn the water into ditches or aqueducts, through which it is conducted to the fields, where it is divided into smaller rills, till it finally disappears. When it is desirable to bring more water upon meadows than is required for saturating the ground, and its escape to fields below is to be avoided, other ditches should be made on the lower sides, to arrest and convey away the surplus water.

The advantages of irrigation are so manifest, that they should never be neglected, when the means for securing them are within economical reach. To determine what economy in this case is, we have to estimate from careful experiment, the equivalent needed in annual dressing with manures, to produce the same amount of grass as would be gained by irrigation; and to offset the cost of the manure, we must reckon the interest on the permanent fixtures of the dam and sluices, and the annual expense of attention and repair.

The quality of grass from irrigated meadows is but slightly inferior to that grown upon dry soils; and for pasturage, it is found that animals do better in dry seasons upon the former, and in wet, upon the latter. In Europe, where the disease is common, sheep are more liable to *rot* upon irrigated and marshy lands, than on such as are free from excessive moisture.

The kind of Soils suited to Irrigation —Light porous soils, and particularly gravels and sands, are the most bene-fitted by irrigation. Tenacious and clay soils are but slight ly improved by it, unless first made porous by underdraining. It is not only important that water be brought on to the ground, but it is almost equally important, that it should pass off immediately after accomplishing the objects sought.

The increase from the application of water, is sometimes fourfold, when the soil, the season and the water are all favorable, and it is seldom less than doubled. Many fields, which in their natural condition, scarcely yield a bite of grass for cattle, when thoroughly irrigated, will give a good growth for years, and without the aid of any manures.

UNDER DRAINING HEAVY AND TENACIOUS CLAY LANDS.

The advancement of agriculture in this country during the few last years, and the high price of farming lands and their products, within convenient distances of our larger markets, fully justify the commencement of an intelligent system of draining on such lands as require it. This system has for many years been introduced and largely practiced in England and Scotland, and has resulted in the most signal success.

The plan first adopted was, to excavate the land in paral-lel lines, at intervals of 16 to 25 feet, and to a depth of two or three feet, forming a slightly-inclined plane on the bottom, which was from three to six inches wide, and gradually en-larging as it approached the surface. The narrowest drains were arched with inverted turf and clay, at such distance from the bottom, as would leave the requisite space for the escape of whatever water might filter through the soil. Others were formed with continuous arched tiles, laid on a *sole*, (a flat tile of the same material,) or on a board placed on the bottom, forming an uninterrupted conductor. Larger ditches were filled with rubble-stone, in some instances brush, to a sufficient depth, and then covered with soil. In all cases, the smaller ones communicated by their outlets, with a large, open drain, which led the water from the field. These drains were always below the reach of the plow, thus leaving the whole surface of the lands free from any obstruc tion to cultivation

Two recent improvements have been introduced, which materially diminish the expense, while they enhance the benefits of the system. They consist in sinking the drain to four feet, and using burnt clay or tile pipes, one and a half to three inches in diameter, and 12 to 18 inches in length, connected together, by allowing the descending end to enter the next below it, as a socket, or by placing the ends close to each other. The slight opening at each joint, with small holes perforating the top of the tiles, is found to be sufficient to admit all the water which falls into the drain; while the increased depth at which the drainage takes place, draws the water from a much greater distance. With the depth indicated, it has been found, that the drains instead of being required once in 16 feet, may be placed at intervals of 35 to 40, and accomplish the object with equal success, *and in less time.* The expense of the former plan, was from $20 to $30 per acre, while the last is only from $12 to $18. For some of the stiffest clays and loosest gravels, these deep drains are not so well suited; as the water scarcely filters through the stiff subsoil of the former, and drains too deeply from the latter.

The advantages of underdraining are numerous and important. They take away all the surplus water which exists in heavy or tenacious soils, and which in wet seasons, is a serious impediment to the successful growth and perfection of vegetation; thus always insuring a full crop, when frequently, not one-fourth of a crop is matured on similar undrained soils. They are susceptible of earlier preparation for the reception of crops in spring, by furnishing a dry, warm soil, which would otherwise not admit of cultivation, except in an advanced stage of the season; thus enabling the farmer to raise a greater variety of products, where only a few were adapted to the soil before; and to these, it gives several weeks additional growth, and an improved quality. The soil is also more porous and friable, and therefore, much more easily tilled. It saves all the trouble and waste of surface drains and open furrows, which require that much of the field be left almost in an unproductive state, to serve as conductors of the surplus water. The rains falling on the convex surfaces of the lands, run off rapidly into the furrows, and not only withhold from the soil those benefits which would result from their absorption, but they carry with them much of the fine soil, which is thus allowed to waste.

Rain water is charged with some of the most important

elements of nutrition for plants, and especially contains con·
siderable proportions of carbonic acid and ammonia. If
these be permitted to percolate through the soil, the roots of
the plants, or in their absence, the elements of the soil itself
absorb and form permanent combinations with them. Air
also holds vegetable food, and it is necessary that this should
penetrate through every portion of the soil where the fibres
of the roots exist. Soils which are saturated with water, do
not admit of any air, unless the small proportion combined
with the water ; and from all such, this vital adjunct of
vegetation is excluded. The porosity of the land thus secured, facilitates the admission and escape of heat, and this
last condition is of the utmost consequence in promoting the
deposition of dews.

The dense mass of saturated soil is impervious to air, and
remains cold and clammy. By draining it below the soil,
the warm rains penetrate the entire mass, and there diffuse
their genial temperature through the roots. Immediately
pressing after these, the warm air rushes in, and supplies its
portion of augmented heat to the land. Porous soils thus
readily imbibe heat, and they as readily part with it ; every
portion of their own surfaces radiating it, when the air in
contact with them is below their own temperature. This
condition is precisely what is adapted to secure the
deposit of the dews, so refreshing, and during a season of
drought, so indispensable to the progress of vegetation.
Dew can only be condensed on surfaces, which are below the
temperature of the surrounding air ; and rapid radiation of
the heat imbibed during the warmth of a summer's day, is
necessary to secure it in profusion for the demands of luxuriant vegetation, in the absence of frequent showers.

An insensible deposit of moisture, precisely analogous to
dew, is constantly going forward in deep, rich, porous soils.
Wherever the air penetrates them, at a higher temperature
than the soils themselves possess, it not only imparts to them
a portion of its exccess of heat, but with it also, so much of
its combined moisture, as its thus ‘lessened capacity for retaining latent heat, compels it to relinquish. To the reflect-
ing mind, imbued with even the first principles of science,
these considerations will be justly deemed of the highest consequence to the rapid and luxuriant growth, and full development of vegetable life.

Another essential benefit derivable from drained lands,
consists in the advantageous use which can be made of the

subsoil plow. If there be no escape for the moisture which
may have settled below the surface, the subsoil plow has
been found injurious rather than beneficial. By loosening
the earth, it admits a larger deposit of water, which requires
a longer time for evaporation and insensible drainage to
discharge. When the water escapes freely, the use of the
subsoil plow is attended with the best results. The earth
being thus pulverized to a much greater depth, and incorporated with the descending pa ticles of vegetable sustenance,
affords an enlarged range for the roots of plants; and in proportion to its extent, furnishes them with additional means of
growth. The farmer thus has a means of augmenting his
soil, and its capacity for production, wholly independent of
increasing his superficial acres; for with many crops, it matters not in the quantity of their production, whether he owns
and cultivates 100 acres of soil, one foot deep, or 200 acres
of soil, half a foot in depth. With the latter, however, he
has to provide twice the capital in the first purchase, is at
twice the cost in fencing, planting and tillage, and pays
twice the taxes. The underdrained and subsoiled fields have
the further advantage, of securing the growth and steady de-
velopement of their crops during a season of drought; as
they derive their moisture from the atmosphere in part, as
before explained, and from greater depths, which are frequently unaffected by the parching heat. This secures to
them a large yield, while all around is parched and withered.*

A more enlarged and general, or what may justly be
termed, a philanthropic view of this system, will readily detect considerations of great moment; in the general heathful-
ness of climate, which would result from the drainage of
large areas, that are now saturated, and in many instances
covered with stagnant waters, and which are suffered to pollute the atmosphere by their pestilential exhalations.

SPRING AND SWAMP DRAINING.

Springs are sometimes discovered, not by a free or open
discharge of their water, but in extensive plats of wet, boggy
lands, which are of no further use than to mire the cattle, and
bear a small quantity of inferior, bog hay. These springs

---

* The experienced reader will sometimes notice the same ideas, repeated under different heads. He must bear in mind, that this work is
intended *for learners*; and that it is of more consequence, thoroughly
to impress their minds with important principles, than to study brevi-
ty in communicating them.

should be sought, at the highest point where the ground appears moistened, and led away to a ravine or rivulet, by a drain, sufficiently deep to prevent the escape of any of the water into the adjacent soil; unless, as it sometimes happens, the position and quality of water are suited to irrigation, when it may be conducted over the field for that purpose.

Swamps and Peat beds occur frequently in a hilly country These are low, level, wet lands, whose constant saturation with water, prevents their cultivation with any useful plants The first object in effecting their improvement, is to find an outlet for the escape of the water, to a depth of three to five feet below the general surface, according to the area to be reclaimed; the greatest depth above specified, being frequently necessary to the effectual drainage, at all times, of an extended field. If the water in the swamp has its origin in numerous springs from the adjoining hills, a ditch should be dug around the entire outer edge, where it meets the ascending land. If the water be derived from a rivulet, a broad ditch should be made as direct as possible from the entrance to its outlet, and deep enough to lead off all the water. If this is found insufficient, additional ones may be made wherever required.

CHAPTER V.

MECHANICAL DIVISION OF SOILS.

SPADING.

AFTER selecting a proper soil and placing it in a proper condition, by manuring, draining, &c., the next most important consideration is, the further preparation of the land for the reception of the seed. In small patches of highly-cultivated land, spading is resorted to, for breaking up and pulverizing the ground more effectually than can be done with the plow. This is the case with many of the market gardens, in the neighborhood of our large cities, and with large portions of Holland, Flanders and other countries of Europe. It is even contended by many intelligent and practical farmers in Great Britain, where labor is about half, and land and agricultural products are nearly twice the average prices with us, that spade-husbandry can be adopted for general tillage crops, with decided advantage to the farmer. However this may be abroad, it is certain it cannot be practised in this country, to any extent, until some very remote period.

There are many important advantages in the deep and minute division of the soil, resulting from the very thorough spading practised by the best gardeners, which we should endeavor to incorporate in every tillage system, with the use of the plow alone. This may be done, and the advantages of spade-husbandry measurably secured, and at one fourth the expense, by the use of the best surface and subsoil plows, if strong teams and skilful plowmen are employed to work them.

PLOWING.

This is the most important of the mechanical operations of the farm. The time, the depth and the manner of plowing must depend on the crops to be raised, the fertility and character of the soil, and other circumstances.

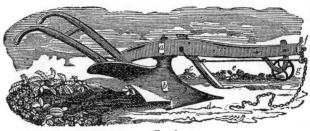

Fig. 1.

The above is a cut of the Eagle plow. This is a good model of a two-horse plow; being easy of draught to the team, and turning a deep, wide furrow, with little effort to the plowman. It is made with a cast-iron mold-board, share and land-side, all in detached parts, which are easily replaced when worn. It has the dial-clevis at the end of the beam, which is an important improvement for controlling the direction of the plow and regulating its depth. Also the draft-rod, which is guided by the clevis, and attached to the beam near the handles. The wheel lessens the draught, facilitates turning the plow at the end of the furrow, and regulates its depth. The cutter or coulter, which is made of wrought iron, with steel edge, divides the sod or earth before the plow, thus lessening the draught, and giving a smoother edge to the furrow.

Plowing Clay lands. — Whenever practicable, these should be plowed in the fall, for planting and sowing the ensuing spring. (For exceptions, see page 18; and for further directions and observations on the subject of plowing, see management of soils preceding, and various subsequent tillage crops.) The tenacity of the soil may thus be temporarily broken up by the winter frosts, its particles more thoroughly separated, and the whole mass reduced to a finer tilth than can possibly be effected in any other manner. A still further and important advantage from this practice ensues, from the attraction existing between the clay and those gases that are furnished from the atmosphere, snow, rains and dews. In consequence of being thus thrown up and coming in contact with them, it seizes upon the ammonia and carbonic and nitric acids, which are in the air, and holds them for the future use of the crops; while their great affinity for manures, effectually prevents the waste of such as are in it. There is an additional benefit sometimes derivable from this practice, in turning over the soil late in the season, and after such worms and insects as are injurious to the crops, have taken up their winter quarters. They are thus thrown out and exposed to the elements, when they are too chilled to seek seclusion again, and are thus destroyed.

The furrows of clay soils, should be turned over so as to lap on the preceding, and lie at an angle of 45°, as illustrated by the following cut :

Fig. 2.

and for this purpose, the depth of the furrow slice should be about two thirds its width. Thus a furrow six inches deep, should be about nine inches wide ; or if eight inches deep, it should be twelve inches wide. This will allow of the furrows lying regularly and evenly, and in the proper position for the drainage of the soil, the free circulation of air, and the most efficient action of frosts, which in this way, have access to every side of them. Land thus thrown up, is found to be finely pulverized after the frosts leave it, and it is comparatively dry and ready for use, some time earlier than such as is not plowed till spring. For sowing, land prepared in this manner requires no additional plowing, but it is better fitted for the reception of seed that it can be by any further operation, unless by a slight harrowing, if the surface is too rough. The different kinds of grain or peas may be dibbled in, or sown broad-cast and covered by the harrow. If sown very early, the grass and clover seeds require no covering, but find their best position in the slight depressions which are everywhere made by the frost, and which the subsequent rains and winds fill up, and cover sufficiently to secure a certain growth. When a field is intended for planting, and is thus plowed in the preceding autumn, in some instances, and especially when the soil is full of vegetable manures, (as from a rich green sward), a single furrow, where the seed is to be dropped, is all that is necessary to be plowed in the spring.

If the land has been previously cultivated, (not in sward), and is designed for planting, a stiff clay is sometimes ridged up, by turning a double furrow, one on each side, and so close together, as partially to lap upon a narrow and unbroken surface; thus leaving the greatest elevations and depressions, which can conveniently be made with the plow. This is shown by the annexed cut.

FIG. 3.

a, a, is surface of the unplowed ground, *b, b, b,* portions of earth not turned by the plow, *c, c, c,* furrows turned over, *d, d, d,* furrows lapped on the preceding. By this means, the frost and air have a greater surface to act upon, than is afforded by thorough plowing, unless it be in a firm sod, which will maintain its position without crumbling. The advantage of a dry surface and early working, are equally secured by this latter method ; and to prepare for planting, the furrows need only to be split, by running a double-mold-board plow through their centre, when they are ready for the reception of the seed.

Plowing sandy or dry Soils.—These require flat plowing, and this may be done when they are either quite wet or dry, but never till wanted for use. By exposure to heat, rains and winds, the light soluble manures are exhaled or washed out, and they receive little compensation for this waste, in any corresponding fertility tley derive from the atmosphere in return.   To insure flat plowing on an old sward, the depth of the furrow should be about one half its width, and the lands or ridges, should be made as wide as possible. This will give more evenness and uniformity of surface, and is an object of importance, where it is to be again laid down as a meadow Some prefer for this purpose, to use the shifting mold-board, side-hill or swivel plow, by all of which names it is known.   This can throw the furrow always in the same direction, and is a right or left hand plow as may be required.

Depth of Plowing.—All cultivated plants are benefitted by a deep, permeable soil, through which their roots can penetrate in search of food ; and although depth of soil is not fully equivalent to its superficial extension, it is evident, that there must be a great increase of product from this cause. For general tillage crops, the depth of soil may be gradually augmented to about twelve inches, with decided advantage. Such as are appropriated to gardens and horticultural purposes, may be deepened fifteen, and even eighteen inches, to the manifest profit of their occupants.   But whatever is the depth of the soil, the plow ought to turn up the entire mass, if within its reach ; and what is beyond it, should be thoroughly broken up by the subsoil plow, and some of it occasionally incorporated with that upon the surface.

The subsoil ought not to be brought out of its bed, except in small quantities.   It should then be exposed to the atmosphere during the fall, winter and spring, or in a summer fallow ; after which, and as a necessary preparation for a crop, it should receive such manr es as are necessary to put it at once into a productive condition.   The depth of the soil can alone determine the depth which the plow should work ; and when the former is too shallow, the gradual deepening of it should be sought, by the use of proper materials for improvement, till the object is fully attained.   Two indifferent soils of opposite characters, as of a stiff clay and sliding sand, sometimes occupy the relation of surface and subsoil towards each other· and when intimately mixed, as they may frequently be by deep plowing alone, and then subjected to the meliorating influence of cultivation, they will frequently produce a soil of great value.

Cross plowing is seldom necessary, except to break up tough sward or tenacious soils, and the former is more effectually subdued by one thorough plowing, in which the sod is so placed, that decomposition will rapidly ensue; and the latter is more certainly pulverized, by incorporating with it such vegetables, and fibrous or unfermented manures, as will produce the same result as the decaying sod. The presence of these in the soil lessens the labor of cultivation, and greatly increases the product.

Subsoil Plowing.—This is a practice of comparatively recent introduction, and it has been attended with signal benefit, from the increase and certainty of the crop. It is performed by subsoil plows, made exclusively for this purpose. The objects to be accomplished, are to loosen the hard earth below the reach of the ordinary plow, and permit the ready escape of the water, which falls upon the surface; the circulation of air ; and a more extended range for the roots of the plants, by which they procure additional nourishment, and secure the crop against drought, by penetrating into the regions of perpetual moisture.

An important additional advantage results from their continued use, in the deepening of the soil. This inevitably follows from opening the subsoil to the meliorating influence of the elements, and from the accumulation of vegetable matter in the roots. These penetrate deeply into the earth, and are left to decay in the bed where they originate. An increased value in the soil has been noticed by observing agriculturists, to the extent, in many instances, of over five per cent. per annum, from the use of this implement. In the rich, compact, deltal lands of Louisiana and elsewhere, the writer has seen the soil made loose, elastic and friable, throughout the whole field, by running a large subsoil plow to its utmost depth, at a distance of four feet between the furrows. The entire mass of soil, seemed to be thoroughly *worked* (mellowed), to the depth which the plow reached, although run at these comparatively remote distances.

When all the circumstances are favorable to the use of the subsoil plow, an increase in the crop of 20 or 30 and sometimes even 50 per cent. has been attributed to its operations. Its maximum influence on stiff soils is reached, only where underdraining has been thoroughly carried out. Its benefits have been more than doubtful, when used in an impervious clay subsoil, where it makes further room for storing up stagnant water; and it is evident they can only aggravate the faults of such subsoils as are naturally too loose and leachy

PLOWS, AND OTHER FARM IMPLEMENTS.

There are plows for almost every situation and soil, in addition to several varieties which are exclusively used for the subsoil. Some are for heavy lands and some for light; some for stony soils, others for such as are full of roots; while several varieties of plows are expressly made, for breaking up the hitherto untilled prairies of the West. Some are adapted to deep and some to shallow plowing; and some are for plowing around a hill and throwing the furrows either up or down, or both ways alternately; others throw the soil on both sides, and are used for furrowing and plowing between the rows of corn or roots. Every farm should be supplied with all that are entirely suited to the various operations required. There is frequently great economy in having a diversity of implements for all the different purposes to be accomplished; and although one of unusual construction may seldom be called for, yet its use for a few days or even for a few hours, may sometimes repay its full cost.

The farmer will find in the best agricultural ware-houses, all the implements necessary to his operations, with such descriptions as will enable him to judge of their merits. Great attention has been bestowed on this subject, by skilful and intelligent persons, and great success has followed their efforts. The United States may safely challenge the world, to exhibit better specimens of farming tools than they now furnish, and their course is still one of improvement. There are numerous competitors for public favor, in every description of farm implements; and an intelligent farmer cannot fail to select such as are best suited to his own situation and purposes.

The best only should be used.—There has been a "penny wise and a pound foolish" policy adopted by many farmers, in their neglect or refusal to supply themselves with good tools to work with. They thus save a few shillings in the first outlay, but frequently lose ten times as much by the use of indifferent ones, from the waste of labor and the inefficiency of their operations. A farmer should estimate the value of his own and his laborer's time, as well as that of his teams, by dollars and cents; and if it requires thirty, ten, or even one per cent. more, to accomplish a given object with one instrument than with another, he should, before buying one of inferior quality, carefully compute the amount his false economy in the purchase will cost him before he has done with it. Poor men, or those who wish to thrive can ill

affoid the *extravagance* of buying inferior tools, at howevei
low a price.   The best are always the cheapest; not those
of high or extravagant finish, or in any respect unnecessari-
ly costly ; but such as are plain and substantial, made on the
best principles and of the most durable materials.   To no
tools do these remarks apply with so much force, as to plows
The improvements in these have been greater than in any
other instruments ; the best saving fully one half the labor
formerly bestowed in accomplishing the same work.

HARROWING.

The object of the harrow, is three fold ; to pulverize the
land, to cover the seed, and to extirpate weeds.   Unless
the land be very light and sandy, the operation should never
be performed for either object, except when sufficiently dry
to allow of its crumbling down into a fine, mellow surface,
under the action of the harrow.   There are several varieties
of harrows in use ; the triangular and the square, both
sometimes hinged and sometimes double ; with long teeth
and with short ones, some thickly set together, and some
wide apart.   For pulverizing firmly-sodded or stiff clay
lands, a heavy, compact harrow is required, with strong
teeth, sufficiently spread ; and for lighter lands, or for covering seed, the more expanded harrow, with numerous small
and thickly-set teeth.   To pulverize well, the harrow should
move as quickly as possible, so as to strike the lumps forci-
bly, and knock them to pieces ; and for this purpose an active
team is required.   When the land sinks much under the
pressure of the horses' feet, l ght animals, as mules or ponies,
are preferable.

THE ROLLER.

This is an important implement for many fields.   It is always useful for pulverizing the soil, which it does by breaking
down such clods and lumps as escape the harrow, and thus
renders the field smooth for the scythe or cradle ; and it is
equally so on meadows, which have become uneven from
the influence of frost, ant-hills, or other causes.   It is ser-
viceable in covering seed, by pressing the earth firmly
around it ; thus securing moisture enough for germination.

But its greatest benefit is with such sandy soils, as are
not sufficiently compact to hold the roots of plants firmly
and retain a suitable moisture.   With these it is invaluable ;
and the proper use of the roller has, in some instances

doubled the product. Its effect is similar to that produced
by the frequent treading in a foot-path; and the observing
farmer will not have failed to notice, the single thread of
thick, green sward, which marks its course over an otherwise almost barren field of sand or loose gravel. The thick-
ly-woven, emerald net-work, that indicates the sheep-walks
on similar soils, is principally due to the same cause. Those
portions of the pasture which have been thus compressed,
will be found to contain a thicker, greener herbage, which
.s earlier in spring and later in autumn, and much more
relished by the animals cropping it, and apparently more
nutritive, than that on other parts of the field.

Rollers are variously constructed. The simplest form is
a single wooden shaft, with gudgeons at each end, which rest
in a square frame, made by fastening four joists together; a
tongue for drawing it being placed in one of its sides. A
box may be attached to this frame, for the purpose of holding stones and weeds picked up in the field, and for weight-
ing the roller according to the work required. The best
rollers, however, are of cast iron, made in sections of about
a foot in length. As many of these may be placed on a
single axle as are required. This division into short sections,
facilitates turning on the axle, either back or forward, and
prevents the unsightly and objectionable furrowing of the
earth, which occurs in turning with the long, solid roller.
Some are made of stone, but these are very liable to break,
and are equally objectionable in turning, as those made
of wood.

The larger the roller, the greater is the surface brought
into contact with the ground, and consequently the more
level it leaves it. To accomplish this object without too
much increase of weight, rollers are frequently constructed
with heads at the ends, and closely covered like a drum. For
dividing compact, clay lumps, or scarifying meadows, they
are sometimes made with large numbers of short, stout,
angular teeth, projecting from the outer or rolling surface,
which penetrate and crush the clods, and tear up and loosen
the old turf and moss of meadows.

THE CULTIVATOR

Has a light frame, in the form of a triangular or wedge-
harrow, with handles behind, like those of a plow, and with
several small iron teeth in the frame, somewhat resembling
a double share plow. The teeth are usually of cast iron, and

when properly made and *chill-hardened*, as is the case with those portions of the plow most liable to wear, they are un-doubtedly the most economical and best. They are rarely made of wrought iron, but more frequently of steel.

They are of various sizes, slightly differing in construc-tion, and are of great utility in stirring the surface of the ground and destroying weeds. By an expanding attachment, they are capable of being adapted to any width of row.

THE CORNPLANTER AND DRILL BARROW

Are useful for dibbling in seeds, and when the surface is mellow, they will open the furrows for the reception of the seed, which it drops, covers, and then rolls the earth firmly over it. The small drills are trundled along like a wheel-barrow, by hand; and the larger, for field sowing, having several fixtures for drilling, are drawn by a horse. These are suited to the smaller seeds. Cornplanters are made to plant corn, beans and peas, which they do at the rate of ten acres per day, and with entire uniformity as to quantity of seed in a hill, depth of covering and distance.

SURFACE OR SHOVEL PLOWS.

These are a cheap and light instrument, much used in England, and to some extent in this country, for paring the stubble and grass roots on the surfaces of old meadows. These are raked together into heaps, and with whatever addition there may be of earth or clay, are burnt, and the ashes and roasted earth scattered over the soil. There is an apparent objection to this practice, from the expulsion of the carbon and nitrogen stored up in the plants; and from the waste of the coarse material of the decaying vegetables, which is so useful in effecting the proper mechanical divi-sions of clay soils. But by a reference to what has been said, on the efficiency of burnt clay or broken bricks, their great utility as fertilizers will be seen. This and the ash of the plants remain; and both are useful in quickening the action of soils, and accelerating those changes, so beneficial to vegetation; and even the re-absorption of the atmosphe-ric gases, it is probable, will more than compensate for those expelled in burning. The effect is further salutary, in de-stroying grubs, insects and their larvæ, and the seeds of noxious weeds.

CHAPTER VI.

THE GRASSES, CLOVERS, MEADOWS AND PASTURES

THE *order* designated by naturalists as *Graminæ*, is one of the largest and most universally diffused in the vegetable kingdom. It is also the most important to man, and to all the different tribes of gramiuiverous animals. It includes not only what are usually cultivated as grasses, but also rice, millet, wheat, rye, barley, oats, maize, sugar-cane, broom-corn, the wild cane and the bamboos, the last sometimes reaching 60 or 80 feet in height. They are invariably characterized as having a cylindrical stem; hollow, or sometimes as in the sugar-cane and bamboos, filled with a pith-like substance; with solid joints and alternate leaves, originating at each joint, and surrounding the stem at their base and forming a sheath upwards, of greater or less extent; and the flowers and seed are protected with a firm, straw-like covering, which is the chaff in the grains and grass seeds, and the husk in Indian corn. They yield large proportions of sugar, starch and fatty matter, besides those peculiarly animal products, albumen and fibrin, not only in the seeds, but also, and especially before the latter are fully matured, in the stems, joints and leaves. These qualities give to them the great value which they possess in agriculture.

Of the grasses cultivated for the use of animals in England, there are said to be no less than 200 varieties; while in the occupied portion of this country, embracing an indefinitely greater variety of latitude, climate and situation, we hardly cultivate twenty. The number and excellence of our natural grasses, are probably unsurpassed in any quarter of the globe, for a similar extent of country; but this is a department of our natural history, hitherto but partially explored, and we are left mostly to conjecture, as to their num bers and comparative quality. Their superior ·ichness and enduringness may be inferred, from the health and thrift o the buffalo, deer and other wild herbivoræ; as well as from the growth and fine condition of our lomestic animals throughout the year, when permitted to range over the woods, and through the natural prairies and bottom lands, where these grasses abound. The writer has seen large droves of the French and Indian ponies come into the settlements about Green Bay and the Fox River in Wisconsin, in the spring, in good working condition, after wintering, entirely on the natural grasses and browse north of lat. 44°.

TIMOTHY, CAT'S TAIL OR HERD'S GRASS (*Phleum pratense.*)

—For cultivation in the northern portion of the United States, I am inclined to place the Timothy first in the list of the grasses. It is indigenous to this country, and flourishes in all soils except such as are wet, too light, dry, or sandy; and it is found in perfection on the rich clays and clay loams, which lie between 38° and 44° north latitude. It is a perennial, easy of cultivation, hardy and of luxuriant growth, and on its favorite soil, yields from one and a half to two tons of hay per acre, at one cutting.

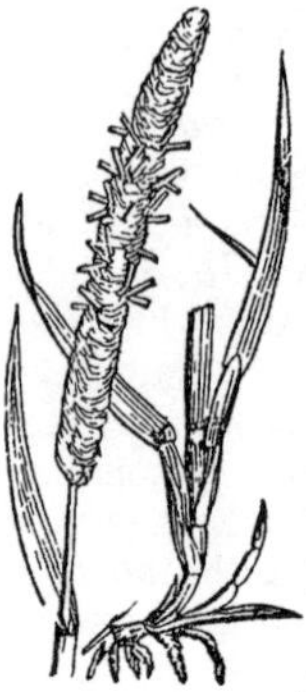

Sinclair estimates its value for hay when in seed, to be double that cut in flower.

Fig. 4.

From its increased value when ripe, it is cut late; and in consequence of the exhaustion from maturing its seed, it produces but little aftermath or rowen. For milch cows or young stock, it should be cut when going into flower, and before the seeds have been developed, as it is then more succulent. It vegetates early in the spring, and when pastured, yields abundantly throughout the season. Both the grass and hay are highly relished by cattle, sheep, and horses; and its nutritive quality, in the opinion of practical men, stands decidedly before any other. It is also a valuable crop for seed, an acre of prime grass yielding from 15 to 25 bushels of clean seed, which is worth in the market from $1 50 to $4 50 per bushel; and the stalks and the chaff that remain, make a useful fodder for most kinds of stock.

It may be sown upon wheat or rye, in August or September or in the spring. When sown either alone, or with other grasses, early in the season, and on a rich soil, it will produce a good crop the same year. From its late ripening, it is not advantageously grown with clover, unless upon heavy clays, which hold back the clover. I have tried it with the northern or mammoth clover, or clay, and found the latter though mostly in full blossom, still pushing out new branches and buds, when the former was fit to cut. The quantity of seed required per acre, depends on the soil and its condition Twelve quarts on a fine mellow tilth, are sufficient, and equal to twice this quantity on a stiff clay. Heavier seed ng than this may be practised with advantage, and especial y, where it is desirable to cover the surface at once with a thick sward.

FIG. 5.   FIG. 6   FIG. 7.

MEADOW FOX TAIL (*Alopecurus pratensis* Fig. 5). This is a favorite grass in England, both for meadows and pasture. It grows early and abundantly, and gives a large quantity of aftermath. It is best suited to a moist soil, bog clay or loam. It is indigenous to the middle States.

SMOOTH STALK MEADOW, GREEN, SPEAR OR JUNE GRASS, the erroneously called BLUE GRASS OF KENTUCKY (*Poa pratensis* Fig. 6), is highly esteemed for hay and pasture. It is indigenous and abounds through the country, but does not appear to reach its highest perfection north of the valley of the Ohio. It is seen in its glory in Kentucky and Tennessee. The seed ripens in June and is self-sown upon the ground, where the succeeding rains give it vitality; and it pushes out its long, rich slender leaves, two feet in height, which in autumn, fall over in thick windrows, matting the whole surface with luscious herbage. Upon these fields, which nave been carefully protected till the other forage is exhausted, the cattle are turned and fatten through the winter. It maintains its freshness and nutritive properties in spite of frost, and the cattle easily reach it through the light snows which fall in that climate A warm, dry, calcareous soil seems to be its natural element, and it flourishes only in a rich upland.

THE ROUGHISH MEADOW GRASS (*P. trivialis* Fig. 7), has the appearance of the *poa pratensis*, but its stalk feels rough to the touch, while the other is smooth. It has the further difference, of preferring moist or wet loams or clay. It yields well and affords good hay and pasture.

FLAT-STALKED MEADOW OR BLUE GRASS (*P. compressa*), is an early dwarfish grass, which abounds in the middle and northern States. It is tenacious of its foothold wherever it intrudes. It possesses little merit as hay, but is valuable for pasture, affording as it does, a close covering to the ground, and yielding much in a small compass.

FIG. 8.   FIG. 9.   FIG. 10.

THE ANNUAL MEADOW GRASS (*P. annua*, Fig. 8).—This grass flourishes in most soils, and in nearly all situations. It affords an early and nutritive herbage, and is relished by all animals. It is perpetually flowering, and affords an abundance of rich seeds. It is hardy and self-propagating, and seldom requires to be sown, but springs up wherever the ground is uncultivated.

NARROW-LEAVED MEADOW (*P. angustifolia*, Fig. 9).— This is an early pasture grass, throwing out a profusion of slender leaves. It flowers late, and before it has reached this point of its maturity, it is liable to rust, which diminishes its value for hay. It is for this reason, as well as its diminutive size. much better adapted to pasture.

RED TOP. HERDS' GRASS, FOUL MEADOW, OR FINE BENT (*Agrostis vulgaris*, Fig. 10), is a hardy, luxuriant grass, loving a very moist soil, and somewhat indifferent as to its texture. The scale of its nutritive properties is put down in the Woburn experiments, at a remarkably low rate, being less than one fourteenth of the value per acre of Timothy in the seed. We think there must be an error in this estimate, as it grows luxuriantly under favorable circumstances, and is relished by cattle. It is seldom cultivated by observing farmers, where the better grasses will grow.

UPRIGHT BENT GRASS, HERDS' GRASS OR FOUL MEADOW (*A. stricta*) is similar to the foregoing, and by some is deemed only a variety.

TALL OAT GRASS (*Avena elatior*) is an early luxuriant grass, growing to the height, sometimes, of five feet. It makes good hay, but is better suited to pasture. It flourishes in a loam or clay soil.

FIG. 12. FIG. 11. FIG. 13.

THE TALL FESCUE (*Festuca elatior*, Fig. 11) would appear by the Woburn experiments, to yield more nutritive matter per acre, when cut in flower, than any other grass cut either in flower or seed. This is a native of the United States, and is best suited to a rich loam. It is not extensively cultivated in this country.

MEADOW FESCUE (*F. pratensis*, Fig. 12) likes a rich boggy soil, bears well and produces an early grass, much relished by cattle, either green, or cured as hay.

SPIKED FESCUE (*F. loleacea* Fig. 13) is adapted to a rich loam, and produces the best of hay and pasture.

THE PURPLE FESCUE (*F. rubra*), SHEEP'S FESCUE (*F. ovina*, Fig. 14) THE HARD FESCUE (*F. duriuscula*

Fig. 15), and THE FLOATING FESCUE (*F. fluitans*, Fig. 16)
are all indigenous to this country, and good pasture grasses.
The two last are good hay grasses, though the former is rather
diminutive.

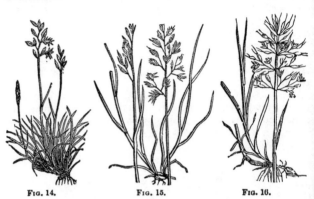

FIG. 14.   FIG. 15.   FIG. 16.

The floating fescue requires to grow in a very wet and
strong clay soil, when it will be found to yield a large burthen
of nutritive forage. The seeds are small but abundant,
sweet and fattening. All fowls are fond of the seed, and all
animals of the seed and herbage.

ORCHARD OR COCK'S FOOT GRASS (*Dactylis glomerata*,
Fig. 17) is indigenous, and for good arable soils, and especially for such as are
shaded, it is one of the most profitable
grasses grown. It should be cut for hay
before it is ripe, as in seeding it becomes
coarse and hard, and is less acceptable to
cattle. It is ready for the scythe with the
clover, and after cutting, it immediately
springs up and furnishes three or four crops
of hay, or constant pasturage throughout
the season. It should be fed closely, to secure a tender, succulent herbage. The
seed is remarkably light, weighing 12 or
FIG. 17.   15 lbs. per bushel. Twenty to thirty pounds
are usually sown upon one acre; yet ten pounds on finely
prepared soils have been known to produce a good sod, over
the entire ground. It flourishes from Maine to Georgia.

AMERICAN OR SWAMP COCK'S FOOT (*D. cynosuroides*) is

an indigenous swamp grass, yielding a large amount of grass **or** hay of inferior quality.

BIENNIAL RYE GRASS, or frequently called, RAY GRASS (*Lolium perenne* or *bienne*, Fig. 18) and ITALIAN RYE GRASS are highly esteemed grasses in Europe. They have been more or less cultivated in this country for many years. They were not successful on their first introduction here, owing to our severe frosts. Recent experiments, however, have shown them to be sufficiently hardy for the middle States; and they are now extensively cultivated in the neighborhood of New York, where they are highly esteemed both for hay and pasturage. On good soils, they yield large returns of valuable forage. It is better to sow early in the spring,

FIG. 18. either by itself or with barley. With oats or wheat, it does not succeed so well. When put in with Timothy and orchard grass, the latter has usurped the place of both the others in a few years. Heavy clay lands are liable to throw out the roots by frost, and thus winter-kill.

FIORIN GRASS (*Agrostis stolonifera*, Fig. 19) has been much lauded in England of late, but it has made little progress in the estimation of American farmers, and probably with sufficient reason. It is a diminutive grass, affording considerable nutriment in a condensed form, and is adapted to a winter pasture. It grows on a moist clay or boggy soil. It is probably on such, and in moist climates only, that it attains its full size, character and value. Many results have been attained with it in England and Ireland, which would seem to commend it, as a valuable forage

FIG. 19. plant, in its appropriate soil and climate. Several of the fiorin family abound in this country, among which is the squitch, couch, or quick grass, which are considered as pests in the cultivated fields

THE SWEET-SCENTED VERNAL GRASS (*Anthoxanthur odoratum*, Fig. 20) is an early and valuable grass, which exhales that delightful perfume so characteristic of much of the eastern meadow hay. It is also a late as well as an early grass, and luxuriates in a dry sandy loam. It affords two, and sometimes three crops in a season.

Poa Alpina (Fig. 21), *Aira cæspitosa* (Fig. 22), *Briza media* (Fig. 23), and the *Agrostis humilis*, and *Agrostis vulgaris*, as well as the *Hard and Sheep's Fescue*, before noticed, are all sweet, pasture grasses, and excellent for lawns. These, and a large variety of other dwarf grasses, abound on our uncultivated uplands, mountains and woodlands, creeping in through the neglect, rather than the care

FIG. 20.

FIG. 21.　　　FIG. 22.　　　FIG. 23.

of the husbandman. They yield a nutritive herbage for the herds and flocks ; and an almost perennial verdure to the landscape, equally grateful to the rustic eye, or a cultivated taste.

RIBBON GRASS (*Phalaris americana*) is the beautiful striped grass, occasionally used for garden borders. It has been highly recommended for swamps, to which, if transplanted, it is alleged that it will supersede all other grasses, and afford a fine quality of hay, of an appearance quite different from the upland growth. The writer tried several experiments, both with the seed and roots, on a clay marsh, but with

out success. Its proper pabulum is probably a rich carbonaceous soil, such as is found in an alluvial swamp or peat
bed.

GAMA GRASS (*Tripsacum dactyloides*) is found growing
spontaneously on a naked sand beach, in Stratford, Ct., and
in other places on our eastern coasts. It has occasionally
been much lauded at the North, where it is a coarse, rough
grass; and it seems generally, to be little prized at the South.
But we have recently, the opinion of some intelligent men
in that section, that it is much relished by stock; as they
frequently eat it so close to the ground, as soon to extirpate
it. We should conclude, therefore, that it is a valuable grass
for some sections of the United States, where the soil and
locality are suited to it.

THE EGYPTIAN, SYRIAN OR GUINEA GRASS (*Sorghum
halpense*, Fig. 24), known by various
other names, is a native of our
southern States, in many of its varieties, although it has been imported from abroad. I have seen it
growing in profusion on Long
Island, Charleston, S. C., and in
southern Mississippi. It grows
like a very slender, miniature cornstalk, from four to six feet high,
with a strong stem, a large grassy

FIG. 24.

leaf, and bears a stately seed-stalk,
tufted with flowrets, which, however, so far as they have
come within my knowledge, do not bear a fully-ripened seed
in this country. That imported from the Mediterranean,
grows with great vigor. Its roots are tuberous, large and
prolific; and equally with the rich, succulent leaves and
stalks, when the latter are young, they are at all times
greedily devoured by stock. Dr. Bachman, of South Carolina, considers it a stock-sustaining plant, far superior to any
other grown at the South. It is difficult to remove when
once embedded in the soil, and the cotton planters look upon
its introduction into their cultivated fields, with unmingled
apprehension.

BERMUDA GRASS (*Cynodon dactylon, Lin.—Paspalum
dactylon, Deccan.*, Fig. 17).—This is considered by **Mr.**
Spalding, an experienced planter in Georgia, who examined
them both critically, from specimens which he raised together, as the *Doub grass of India*, so much commended
5*

by Sir William Jones, and so highly prized by the Brah-
mins. It is by the agriculturists of the South, deemed an
invaluable grass, yielding four or five tons per acre on good
meadow. Mr. Affleck, of Mississippi, states the yield of three

Fig 25.

cuttings, at five to six tons per acre on common meadow,
that it loses only 50 per cent. of its weight in drying, and is
consequently the hardest grass to cut. It is one of the most
nutritive grasses known, and is of great value to the river
planter. It loves a warm and moist, but not wet soil.

Crab Grass is considered (unjustly as I think) a pest by
the cotton-planters, for equally, perhaps, with the Bermuda,
it is a rich and nutritious grass. It comes up after the crops
are *laid by* (received their last plowing and hoeing), and
grows rapidly as the cotton or corn matures and dries; and
by the time they are ready to remove from the field, has
frequently attained so large a growth, as to afford a crop of
hay. Even considered as a fertilizer alone, it is a valuable
assistant to the planter. When the corn or cotton is young,
the ground requires working to an extent sufficient to
keep down this grass, solely with a reference to preserving
its porosity—its dew-condensing, dew-absorbing properties. When the crop is sufficiently matured to need no further care, the grass shoots forward rapidly, and absorbs
largely from the floating elements of the air.

Winter Grass is known on the low, moist fertile soils of
Mississippi and adjoining States. It springs up in the autumn, grows all winter, and seeds in the spring. It fattens
all animals that feed upon it.

The Muskeet Grass, found growing on the plains of Mexico and Texas, is considered one of the best of the indigenous
grasses. I have seen it growing on the plantations of Louisiana, where it has been successfully transplanted.

Grama (*La Grama*, or the *grass of grasses*) is held in
the highest estimation by the Mexicans. It attains a medium height, and is deemed the most nutritious of the natural
grasses in our southwestern, frontier prairies, in California,
and parts of Mexico. It grows on dry, hard, gravelly soils,
on side hills, the swells of the prairies, and the gentle eleva-

tions in the valleys. The principal value is found in the numerous seeds, which are retained in the pods with great tenacity, long after they are ripe, serving as a luxurious food for all the graniverous beasts and fowls of the regions where it is grown.

The Buffalo Grass is found intermixed with the Grama and seldom grows more than a few inches in height. It forms a thick, soft herbage, on which the traveller walks with ease, and reposes when weary, with delight. It yields a rich sustenance to countless herds of wild horses and cattle, buffaloes, deer and antelopes.

Tornillo or Screw Grass.—This grows in great profusion in the region of the two last grasses; but is most conspicuous on the table lands, and between the rivers and creeks, the tall grass of the lower levels, giving place to it as the surface ascends. It is taller than the buffalo, with broader leaves. It bears a seed-stock eight or ten inches high, surmounted by a spiral-shaped pod, an inch long and one-fourth of an inch in diameter, which contains ten or twelve roundish, flattened seeds. The herbage is not relished by animals, but the ripened seeds yield a food of great richness, on which innumerable herds of wild cattle fatten for slaughter. Horses, mules and most other animals and fowls subsist upon it.—(*Dr. Lyman.*)

The Prairie Grasses abound in the western prairies, and are of great variety, according to the latitude and circumstances under which they are found. They afford large supplies of nutritive food both as pasturage and hay. They possess different merits for stock, but as a general rule, they are coarse when they have reached maturity, and are easily injured by the early frosts of autumn. Some of the leguminosæ or wild pea vines, which are frequently found among them, yield the richest herbage. We are not aware that any of these grasses have been cultivated with success.

The Pony Grass (Fig 26) may be mentioned, as one of the

Fɪɢ. 26.

best of the winter grasses in our western States. It grows in close, thick, elevated tufts, and continues green all winter. It is easily detected under the snow by animals, from the little hommocs which everywhere indent its surface.

The Wild Rice which lines the still, shallow waters of

the streams and small inland lakes of many of the western States, affords a palatable forage when green, or if early cut and dried ; and the grain, which is produced in great profusion, is an exhaustless store to the Indians, who push into the thickest of it, and bending over the ripe heads, with two or three strokes of the paddle on the dry stalks, rattle the grain into their light canoes. The wild ducks, geese and swans, which yet frequent those waters, fatten on this grain throughout the fall and winter.

TUSSAC GRASS (*Dactylis cespitosa*) is a luxuriant, salt-marsh grass, growing in large tufts, and is found in perfection on its native soil, in the Falkland Islands, between 51° and 52° South, and about 8° east of the Straits of Magellan. Capt. Ross describes it, as " the gold and glory of those islands. Every animal feeds upon it with avidity, and fattens in a short time. The blades are about six feet long, and from 200 to 300 shoots spring from a single plant. About four inches of the root eats like the mountain cabbage. It loves a rank, wet peat bog, with the sea spray over it." Governor Hood of those islands says, " to cultivate the tussac, I would recommend that the seed be sown in patches, just below the surface of the ground, and at distances of about two feet apart, and afterwards weeded out, as it grows very luxuriantly, and to the height of six or seven feet. It should not be grazed, but reaped or cut in bundles. If cut, it quickly shoots up, but is injured by grazing, particularly by pigs, who tear it up to get at the sweet nutty root."

ARUNDO GRASS (*Arundo alopecurus*).—Mr. Hooker from the same islands says : " another grass, however, far more abundant and universally distributed over the whole country, scarcely yields in its nutritious qualities to the tussac ; I mean the Arundo alopecurus, which covers every peat bog with a dense and rich clothing of green in summer, and a pale, yellow, good hay in the winter season. This hay, though formed by nature without being mown and dried, keeps those cattle which have not access to the former grass, in excellent condition. No bog, however rank, seems too bad for this plant to luxuriate in ; and as we remarked during our survey of Port William, although the soil on the quartz districts was very unprolific in many good grasses, which flourish on the clay slate, and generally speaking, ot the worst description, still the Arundo did not appear to feel the change ; nor did the cattle fail to eat down large tracts of this pasturage."

I have purposely devoted several pages to the description

of such new grasses as are indigenous to this continent, and which, by their superior value in their native localities, would seem to commend themselves to a thorough trial in similar situations elsewhere. There are doubtless, others of great merit, which experiment hereafter, will demonstrate to be of singular benefit to the American farmer. Most of these, yet remain to be classified by the botanists; and what is of much more utility, to be thoroughly tested by the crucibles of intelligent chemists and the experiments of enlightened agriculturists, to determine their absolute and relative value for economical purposes. The subject of grasses has been but slightly investigated in this country, in comparison with its immense importance; and for this reason, with few exceptions, we are at a loss for the true comparative value, of the foreign and indigenous grasses, to American husbandry.

As an instance of the want of a well-established character to some of our most generally-cultivated grasses, we quote the opinions of Dr. Muhlenburgh, of Pa., who has written ably on the subject; and the late John Taylor, a distinguished agriculturist of Virginia, both of whom place the *tall oat grass (Avena elatior)* at the head of the grasses; yet from the investigations made at Woburn, it appears among the poorest in the amount of nutritive matter yielded per acre. Dr. Darlington, also of Pennsylvania, does not mention it, but gives the following, as comprehending " those species which are considered of chief value in our meadows and pastures, naming them in what I consider the order of their excellence : 1. Meadow or green grass *(Poa pratensis)* 2. Timothy *(Phleum pratense)*. 3. Orchard grass *(Dactylis glomerata.)* 4. Meadow fescue *(Festuca pratensis)*. 5. Blue grass *(Poa compressa)*. 6. Ray grass *(Lolium perenne.)* 7. Red top *(Agrostis vulgaris)*. 8. Sweet-scented vernal grass *(Anthoxanthum odoratum)*."

The *Sweet-Scented, Soft Grass,* or *Holy Grass, (Holcus odoratus)*, according to the Woburn table, is next to the *tall fescue* and *Timothy* in point of nutritive matter to the acre, when cut in seed, and it is placed as far in advance of all others, in the value of its aftermath; yet scarcely any other authority mentions it with commendation.

Without relying on these experiments, as an unerring guide for the American farmer, we append the table on the two following pages, as the fullest and most correct we have on the subject, and as affording a useful reference to some of the leading and most desirable of the English grasses many of which are more or less cultivated in this country

Table of the Comparative Product and Value of Grasses, as Experimented on at Woburn, by Mr. GEORGE SINCLAIR, under the direction of the DUKE OF BEDFORD.

BOTANIC AND ENGLISH NAMES OF PERENNIAL GRASSES.	Height in wild state in inches	Soil employed.	When weigh'd	Wt. per acre when green.	Wt. per acre when dried.	Loss in drying.	64 drms gave nutritive mat'r (d / gr)	Nutritive matter in one acre.	When in flower.	When in seed.	Proportionate value of the grass in seed? (F'r : Sd)	Gene'al character
Anthoxanthum odoratum*—Sweet-scented vernal grass.	12	Sandy loam.	In flower. / In seed. / L. Math	7827 / 6125 / 6806	2103 / 1837	5723 / 4287	1 0 / 3 1	122 / 311 / 238	April 29	June 21.	4 to 13	Early pasture grass.†
Holcus odoratus. Host.—Sweet-scented soft grass.	14	Rich sand loam.	In flower. / In seed. / L. Math.	9528 / 27225 / 17015	2441 / 9528	7087 / 17696	4 1 / 5 1 / 4 1	610 / 2933 / 1129	April 29	June 25.	17 to 21	The most nutritive, early grass.†
Alopecurus pratensis—Meadow fox-tail.	24	Clay loam.	In flower. / In seed. / L. Math.	24418 / 12931 / 88167	6125 / 5819	14293 / 7111	1 1 / 1 1 / 1 0	270 / 461 / 255	May 30.	June 24.	9 to 6	One of the best meadow grasses.‡
Poa pratensis*—Smooth-stalked meadow grass.	18	Bog earth & clay.	In flower. / In seed. / L. Math.	10209 / 8607 / 4083	2871 / 3403	7337 / 5104	1 3 / 2 1 / 1 1	229 / 199 / 111	May 30.	July 14.	……	Good early hay grass.‡
Avena pubescens*—Downy oat grass.	18	Rich sand loam.	In flower. / In seed. / L. Math.	16654 / 6806 / 6806	5570 / 1361	9788 / 5445	1 2 / 2 0 / 2 0	366 / 212 / 212	June 15.	July 8.	6 to 8	Good pasture grass.†
Poa trivialis*—Roughish meadow grass.	20	Manu-red lgt. loam.	In flower. / In seed. / L. Math.	7487 / 7827 / 4764	2246 / 3522	5240 / 4304	2 0 / 2 3 / 3 0	233 / 336 / 223	June 15.	July 10.	8 to 11	Good on rich moist soils.†
Agrostis stricta*—Upright bent grass.	9	Bog soil.	In flower. / In seed.	7486 / 4764	2713 / 1310	4772 / 3454	1 2	446 / 47	July 28.	Aug. 30	8 to 5	‡
Festuca rubra*—Purple fescue grass.	12	Light sand.	In flower. / In seed. / L. Math.	10209 / 10890 / 30403	355. / 4900	6851 / 5939	2 0 / 2 2	239 / 340 / 79	June 20.	July 10.	6 to 8	Good long gr.
Festuca ovina—Sheep's fescue grass.	6	Light sand.	In flower / in seed / Math	5445				127	June 24	July 10		Good long gr

Grass	Depth of soil	Soil	State						Flowers	Cut for hay	Produce	Remarks
....foot grass.....	24	sand soil	In seed. / L. Math.	26544 / 11910	13272 / 7810	13272 / 10666	3 2 / 5 0	145 / 1430	June 24.	July 14.	5 to 7	....ductive grs. but coarse.‡
Poa angustifolia—Narrow-leaved meadow grass......	24	Brown loam	In flower. / In seed.	18376 / 9528	7810 / 3811	10666 / 5717	5 0 / 5 1	1430 / 701	June 28.	July 16.	3 to 2	Excellent hay grass.‡
Trifolium pratense—Red clover.....	....	T. clay	In seed.	49005	12251	3675	2 2	1914	July 18.	Aug. 6.		
Medicago sativa—Lucern......	36	Cl.lo'm.	In seed.	70786 / 28314	7659 / 4271	5208	1 2	7659	July 18.	Aug. 6.		Best for soil'g.‡ ††
Hedysarum onobrichis—Sainfoin.....	....	Cl.lo'm.	In seed.	8848	3539	5208	2 2	345	July 18.	Aug. 6.		
Festuca duriuscula*—Hard fescue grass.....	12	Sandy loam	In flower. / In seed. / L. Math.	18376 / 19075 / 10209	8969 / 8575	10116 / 10481	3 2 / 1 2 / 1 1	1004 / 446 / 199	July 1.	July 20.	14 to 6	Good for hay or pasture.‡
Festuca pratensis—Meadow fescue grass......	30	Bog soil	In flower. / In seed.	13612 / 19057	6465 / 7623	7046 / 11435	4 2 / 1 2	957 / 446	July 1.	July 20.	18 to 6	Excellent for early hay.‡
Lolium perenne—Perennial rye grass......	24	Rich brown loam	In flower. / In seed. / L. Math.	7827 / 4492 / 3403	3222 / 4492	4494 / 10481	2 3 / 1 0	305 / 643 / 53	July 1.	July 20.	10 to 11	Generally esteemed.†
Festuca loliacea—Spiked fescue grass......	36	Rich brown loam	In flower. / In seed. / L. Math.	16335 / 10890 / 3403	7146 / 4492	9189 / 6397	3 0 / 3 1	765 / 553 / 166	July 1.	July 28.	13 to 12	Most valuable for hay and pasture.‡
Avena elatior—Tall oat grass......	50	Brown loam	In seed. / L. Math.	16335 / 13612	5717	10617	1 0 / 1 1	265 / 265	July 6.	July 28.		Good long gr.†
Festuca elatior*—Tall fescue grass......	36	Black rich loam	In flower. / In seed. / L. Math.	51046 / 51046 / 15664	17866 / 17866	33180 / 33180	5 0 / 3 0 / 4 0	3988 / 2392 / 978	June 28.	July 16.	20 to 12	Ex. mead. gr.‡
Festuca fluitans*—Floating fescue grass......	18	St. clay	In seed.	13612	4083	9528	1 3	372	July 14.	Aug. 12.		Aquatic grass.
Holcus lanatus*—Meadow soft grass......	24	St. clay loam	In flower. / In seed.	19057 / 19957	6661 / 3811	12395 / 15246	4 0 / 2 3	1191 / 898	July 14.	July 26.	12 to 11	Early and productive.‡
Poa fertilis—Fertile meadow grass......	20	Cl.lo'm.	In flower.	14973	7861	7111	4 2	1042	July 14.	July 28.		An early gr.
Phleum pratense—Meadow cat's tail grass....	24	Clay loam	In flower. / In seed. / L. Math.	40837 / 40837 / 9528	17355 / 19397	23481 / 21439	3 3 / 2 0 / 2 0	1595 / 3669 / 297	July 16.	July 30.	10 to 23	Ex. for hay.†
Avena flavescens—Yellow oat grass......	18	Clay loam	In flower. / In seed. / L. Math.	8167 / 12251 / 4083	2558 / 4900	5308 / 7350	3 3 / 2 1	478 / 030 / 79	July 24.	Aug. 15.	15 to 9	Valuable gr.‡
Agrostis vulgaris—Fine bent grass......	18	S. Soil	In seed.	9528	4764	4764	1 2	251	July 34.	Aug. 20.		An early gr.

* Natives of the United States. † Best cut in seed. ‡ Best cut in flower. § A perennial

SOWING GRASS SEEDS.

As a general rule, grass seeds do best when sown early in the spring, on a fine tilth or mellow soil. If this is done while the frost is leaving the ground, no harrowing will be necessary, as the spring rains wash the seed into the honeycomb left by the frost, and secure to it an early germination. They are also successfully sown in August or September, when the fall rains will generally give them sufficient growth, to withstand the effects of the succeeding winter, if the land be free from standing or surface water.

It has recently been the practice of many judicious farmers, to renovate their old, worn-out meadows, by giving them a coating of unfermented manure, and then turn the sod completely over. On the surface thus plowed, a dressing of well-rotted manure or compost with ashes, is spread and thoroughly harrowed lengthwise of the furrows. The seed is then sown and slightly harrowed in, and the decomposing manure and the stubble and roots of the sod give an immediate and luxuriant growth.

Grass seed is generally sown with the white grains, wheat, rye and oats; but if the grass be sown alone and sufficiently thick, the young plants will exclude the weeds, and occupy the soil as profitably as can be done with the grain. Though the moisture and shade, which are secured by the presence of the grain, are sometimes an advantage to the grass seed, yet it often fails when thus sown, from the absence, perhaps, of sufficient sun and air; or more probably, from the exhausting crop of grain, which precedes it. We do not sufficiently appreciate the violation of one of the essential principles of rotation in this practice, as the grass is of the same class of plants as the grain, which has just been taken from the field. When followed by clover, this objection fails.

There is usually a great deficiency of grass seed sown, when permanent meadows or pastures are required. The English method is, to mix together and sow on a single acre without any grain, two to four bushels of various seeds, which are the best adapted to the purpose. A quick and full growth rapidly covers the surface with a rich herbage, frequently surpassing in value, that of the best natural pastures or meadows.

LANDS THAT SHOULD BE KEPT IN PERPETUAL GRASS,

Are such as are frequently under water, as salt and fresh

water meadows; such as are liable to overflow, as the rich bottom or interval lands upon a river bank; heavy, tenacious clays, and mountain or steep hill-side land, which is peculiarly liable to wash from rains. The low, bottom lands generally receive one or more annual dressings from the over flowing waters. The fertilizing matters thus deposited, are converted into hay, and become a reliable source for increasing the muck heap for other parts of the farm, without demanding any thing in return. The thick sward of nutritious grasses, which nature has so lavishly supplied to them, is an effectual protection against abrasion and waste from the overflowing water, while the crop, if at any time submerged, can receive comparatively little injury. If plowed and the fine loose earth is exposed to a sweeping current, much of the soil and all the crop may be lost.

Strong clay lands cannot be properly worked without much labor, unless when under-drained and well filled with manure, and they seldom exist in the former condition in this country. Yet these soils, next to the fertile, self-sustaining, bottom lands, are the most profitable for the various grasses. When thus appropriated, immediately after clearing off the native growth of wood, the fine vegetable mold at the surface, aided by the magazine of supplies contained in the clay below, gives to them the most certain and permanent growth. If once plowed, this mold is turned under, and the intractable clay takes its place on the surface; and lacking those peculiarities of color, texture and chemical composition, which we have before shown, are essential to the most successful vegetation, the grass is thin, and for years, comparatively unproductive. When necessary to break up such lands, they ought to be thoroughly manured, evenly laid down, and heavily seeded to grass; and if any deficiency of seed or growth is manifested, they should receive an addition of seed, with a compost dressing.

The injury to plowing steep side-hills, is sufficiently apparent, as not only the soluble matters, but many of the finer particles of the soil, are washed out and carried far beyond reach. Such lands should be kept in permanent pasture, if not suitable for mowing. If fed off by sheep, they drop most of their manure on the higher points, which is partially washed down and sustains the fertility of every part. There is still another class of lands that should not be broken up for meadows. These are such as are filled with small stones from the surface of which they have once been cleared; but which plowing and harrowing will again bring to it, and here they will remain, a perpetual annoyance to the mower, unless removed at no little trouble and expense.

MEANS OF RENOVATING PERMANENT MEADOWS AND PASTURES.

The general theory adopted in regard to pasture lands, is, that they are manured sufficiently by the animals feeding on them. This opinion is only partially correct. Pastures wear out less than other lands; but when milch cows and working animals are fed upon them, they carry off much of the produce of the soil, which is never again returned to it. Even the wool and carcass of sheep, with the ordinary escape of the salts by the washing of the rains, will, after a long time, impoverish the land. How much more rapidly then, if much of the manure and all the milk, which is rich in all the elements of plants, is daily carried from the soil. To such an extent have the permanent, clay pastures of Cheshire (England) been impoverished, that it has been found necessary to manure them with crushed bones, which at once brought up their value more than 100 per cent. There is much phosphate of lime in milk, and bones, which are mostly of the same material, are the best manure that could be used for dairy pastures. Wool contains a large proportion of sulphur, and sulphate of lime (gypsum) is therefore a proper manure for sheep pastures. Whatever has a tendency to develope vegetation, will generally accomplish the object by yielding all the needful properties. Ashes and salt are of the highest value for pasture lands, and with the addition in some instances, of lime, bones and gypsum, are all that would ever be necessary for permanent pastures. From the peculiar action of these, instead of growing poorer, *pastures may become richer through every successive year.*

Permanent meadow lands, if constantly cropped without manures, may be exhausted with much greater rapidity than pastures, though this depreciation is much more gradual than with tillage lands. There is no greater mistake than to suppose they will keep in condition, by taking off one annual crop only, and either pasturing the aftermath, or leaving it to decay on the ground. By recurring to the table of the ash of plants, page 35, it will be seen, that the analysis of hay there given, shows over five per cent., while dried clover yields from seven to nine per cent. of earthy matter. Every particle of this is essential to the success of the plant ; and yet, if the land produces at the rate of two tons per acre, the salts are taken out of it, to the amount of upwards of 300 lbs. per annum. No soils but such as are periodically flooded with enriching waters, can long suffer such a drain with impunity. *They must be renewed with the proper manures, or barrenness will ensue.* Ashes, lime, bones, and gypsum (the latter especially to be applied to clovers, its good effects not being so marked on the grasses), are essential to maintain fertility ; and to insure the greatest product, animal or vegetable manures must also be added.

The proper manner of applying manure, is by mixing in a compost and scattering it over the surface, when the grass is just commencing a vigorous growth in spring, or simultaneously with the first rains after mowing. The growing vegetation soon buries the manure under its thick foliage, and the refreshing showers wash its soluble portions into the roots ; and even the gases that would otherwise escape, are immediately absorbed by the dense leaves and stalks, which everywhere surround it. When scattered broadcast, under such circumstances, the loss of manure is trifling, even in a state of active decomposition.

Pasturing Meadows.—It is an established principle with some, that close feeding, as often at least, as once a year, is essential to the permanent productiveness of all meadows. There is certainly no objection to feeding them soon after being mown, and while the ground is dry and the sod firm. The roots of the grass are rather benefitted than injured by the browsing, and the land is improved by the droppings from the cattle. But they should never be pastured in early spring or late autumn. It is economy to purchase hay at any price, rather than to spring-pasture meadows, or feed them too late.

Rotation on Grass Lands.—Most soils admit of a profitable rotation or change of crops ; and where this is the case, it is generally better to allow grasses to make up one of the items in this rotation. Where these are successfully grown in permanent meadows, this change or breaking up is less to be sought on their own account, than for the other crops, which do better for having a rich, fresh turf to revel in. Thus, potatoes are sounder and better, and yield more on turf than on old plowed ground; and the grain crops are generally more certain and abundant on this, than on other lands But many of the light soils retain the grasses only for a short time. These should be placed in a rotation, which never assigns more than two years to grass.

Time for cutting Grass.—This must depend on the kinds of grass cultivated. We have seen, that Timothy affords nearly double the quantity of nutriment, if cut after the seed has formed, instead of while in flower, and it is then much more relished by horses and a portion of the stock. Timothy therefore should never be cut for them, until after the seed has filled. The proper time for harvesting, is between the milk and dough state, when it will nearly ripen after cutting. Orchard grass, on the other hand, although possessing two-sevenths more nutritive value for hay in the seed, yet as it is more tender, and much preferred by stock when cut in flower, and as it continues to grow rapidly afterwards, should be always cut at that time. Even a few days will make an important difference in the value of grass, when cut for hay. The kind of grass, and the stock to which it is to be fed, cannot, therefore, be too closely noted, to detect the precise moment when the grass will best subserve the purpose for which it is intended.

Curing Grass.—Many farmers do not consider the scorching effects of our cloudless July suns, and the consequence is, that hay is too much dried in this country. Unless the grass be very thick and heavy, it will generally cure sufficiently, when exposed in the swath for two days. When shook or stirred out, it should not remain in this condition beyond the first day, or it will thus lose much of its nutritive juices; nor should dew or rain be permitted to fall upon it, unless in cocks. It is better, after partially drying, to expose it for three or four days in this way, and as soon as properly cured, place it under cover. It is a good practice, to salt hay when put up, as it is thus secured against damage from occasional greenness; and there is no waste of the salt, as it serves the double object after curing the hay, of furnishing salt to the cattle and the manure heap.

There is a loss of available, nutritive matter, in the ordinary mode of curing hay, which is obvious to every careful feeder. This is conspicuously evident, in the diminished quantity of milk yielded by cows, when taken from the pasture and put upon the hay made from grass, similar to that before consumed. To what this difference is owing, is not yet fully ascertained; but it is undoubtedly the result of several causes combined.

The tender, succulent grass, in the process of excessive

drying, is partially converted into woody fibre, a form in some degree, equally removed from the nutritive properties of the green herbage, as slabs or saw dust from the life-sustaining principles, yielded by fresh, young boughs and twigs. When there is mismanagement in the curing process, resulting in fermentation, the saccharine matter, so abundant in the juices of good grass, and so essential to some of the constituents of milk, is converted into alcohol and carbonic acid, both of which rapidly escape, and would be useless to the animal if retained. A series of careful experiments has been made, which showed the important fact, that a cow, thriving on 100 to 120 lbs. grass per day, required nine pounds of barley or malt in addition to this quantity, when converted into hay. This is stated as illustrating a general principle, without assigning to it any definite or uniform ratio of deterioration, which varies with every variety of grass, and the period and manner of curing.

THE CLOVERS,

Sometimes, improperly called grasses, are botanically arranged in the order, *leguminosæ*, under the same head with the bean, pea, locust and vetches. More than 160 species of clover have been detected by naturalists. Their properties and characteristics are totally unlike the grasses, with which they agree, only in their contributing in a similar manner to the support of farm-stock. There are many varieties cultivated abroad, but the attention of farmers in this country, has been limited to a very few.

THE COMMON RED OR NORTHERN CLOVER, (*Trifolium pratense,*) a biennial, and occasionally, on calcareous soils, a triennial, is the species most generally in use in the United States. This is a hardy, easily-cultivated variety, growing luxuriantly on every properly-drained soil, of sufficient strength to afford it nutriment. It has numerous, strong, well-developed stems, branching outwardly and vertically from a single seed, each bearing broad, thick leaves, which are surmounted by a large, reddish, or purple flower. By the analysis of Davy, the whole plant yields an amount of nutritive matter, fully equal to any other of the clovers.

Mode of Cultivation.—Clover may be sown broadcast, either in August or September, or early in the spring, with most of the cereal grains, or the cultivated grasses; or it may profitably constitute a crop by itself. The quantity of seed required per acre, depends on the kind of soil. On well prepared loams, ten or twelve pounds of good seed will frequently give a full covering to the land, while on clay twelve to sixteen pounds are necessary per acre. When sown with the grasses, six on the first, and eight to twelve pounds on the last soil will suffice. An additional amount of seed, as with the grasses, will give a finer quality of hay, in consequence of multiplying the number of stalks; and for this purpose, as well as to insure it on every spot of the field, it should always be liberally sown. The covering, like that of grass seeds, should be of the slightest kind; and when sown very early in the spring, or on well pulverized grounds and followed by rains, it will germinate freely without harrowing.

After the leaves are developed in the spring, an application of gypsum should be made by sowing broadcast, at the rate of one to three or four bushels per acre. The effect of this on clover, is singularly great, and it seems to be augmented by applying it on the leaves. This may perhaps be accounted for, in the fact, that besides its other uses, gypsum yields a considerable proportion both of its sulphuric acid and lime to the plant, and thus constitutes a direct food. The influence of gypsum is almost incredible, in bringing up the clovers on fields where they were hardly discernible before. This may be witnessed in almost every soil where gypsum has any effect. By sowing a quantity over the grass plat containing either the seeds or plants of the clover, however thin or meagre they may be, an immediate and luxuriant growth distinguishes the spot which has received it, from all the surrounding field.

Bones are invaluable manure for the clovers. The table of the ashes (page 35), shows the great quantity of lime and phosporic acid (the leading elements of bones), which the clovers contain, in comparison with the rye grass, which is a type of the other grasses. Thus, the red clover has about four times as much lime, twenty-six times as much phosphoric acid, more soda and sulphuric acid, and nearly twice and a half as much potash as the grass. The white clover has about four times the potash; the lucern, nearly seven times the lime, and fifty-two times the sulphuric acid, contained in the grass.

Such are the various demands of plants, and the necessity of providing each with its specific food. And hence, the advantage of cultivating a variety of grasses and clovers on the same spot. Each, it is true, draws its nutriment from the same elements, but in such unlike proportions, that when they cease to yield adequate support to one, the soil may still be rich in those which will give luxuriant growth to others. Thus, two or more of the forage plants, when growing together, may each yield a large crop, swelling the aggregate product far beyond what would be realized in the separate cultivation of either. This is a conspicuous and satisfactory illustration of the utility of good husbandry, as shown in the cultivation of the mixed grasses and forage.

Time for cutting and mode of curing Clover.—Clover should be cut after having fully blossomed and assumed a brownish hue. By close cutting, more forage is secured, and the clover afterwards springs up more rapidly and evenly. The swath unless very heavy, ought never to be stirred open, but allowed to wilt on the top. It may then be carefully turned over, and when thus partially cured, placed in high slender cocks, and remain till sufficiently dry to remove into the barn. Those who are very careful in curing their hay, provide cheap cotton covers (tarpaulins are better), which are thrown over the cocks when exposed to the rain, the corners of which are weighted, to prevent being blown up by the wind. The long exposure of clover to the weather, when thus cured, renders this precaution peculiarly desirable. The clover may be housed in a much greener state, by spreading evenly over it in the mow, from ten to twenty quarts of salt per ton. Some add a bushel, but this is more than is either necessary for the clover, or judicious for the stock consuming it; as the purgative effects of too much salt, induce a wasteful consumption of the forage. A mixture of alternate layers of dry straw with the clover, by absorbing its juices, answers the same purpose, while it materially improves the flavor of the straw for fodder.

After-management of Clover fields.—The second crop of clover may be either saved for seed, mown, pastured, or turned under for manure. As this is a biennial when allowed to ripen, the stocks generally die after the second year; and the crop is only partially sustained afterwards, by the seed which may have germinated the second year from the first sowing, or from such as has been shed upon the surface, from the seed matured on the ground. The maximum of benefit derivable to the soil, in the manure of the stubble and roots, is attained the second year; as we have seen that the dried roots of the clover at that time, are sometimes in the proportion of 56 for every 100 pounds of clover hay produced from them in two years. But the ground is then so full of the roots, as

to check further accumulation. This is then the proper time
for plowing up the field, and renewing again its accustomed
round of crops. If desirable, the clover may be imperfectly
sustained on some soils, for a few years, by the addition of
gypsum, bone-dust, ashes and other manures, which will develop and mature the ripened seeds; but the greater tenacity of other plants and grasses, will soon reduce it to a
minor product in the field.

Complaint is sometimes made among farmers in England,
whose fields have been often in this crop, that their land is
clover sick. This arises, simply, from the exhaustion of the
land of some of those principles peculiar to clover, which are
needed to prepare them again for bearing good crops. Ro-
tations and judicious manuring are the only remedies for this,
or similar deficiencies with other crops.

Importance of the Clovers.—The great value of the different clovers as forage, was well known to the ancients.
They were extensively cultivated by the early Romans, and
since that period, they have been extended throughout a large
part of Europe. They were not introduced into Great Britain till the 16th century, but have since constituted a profitable branch of its husbandry. Their importance has long
been acknowledged in the United States. The nutritive
matter, although relatively less than from some of the grasses,
is yet, in the amount per acre, fully equal to the average
of any other forage crop, which is produced at the same expense. It is easily and cheaply raised; it is liable to few
or no casualties or insect enemies in this country; and its
long tap roots are powerful auxiliaries in the division
and improvement of soils. Its broad, succulent leaves
derive a large portion of their nutriment from the atmosphere;
and while it affords a product equal to the best grasses, it
draws a large part of it from the common store-house of
nature, without subjecting the farmer to the expense of providing it in his manures.

It is as a fertilizer, however, that it is so decidedly superior to other crops. In addition to the advantages before
enumerated, the facility and economy of its cultivation, the
great amount yielded; and lastly, the convenient form it offers
for covering with the plow, contribute to place it far above
any other species of vegetation for this purpose. All the
grains and roots do well after clover; and wheat especially,
which follows it, is more generally free from disease than
when sown with any other manure. The introduction of

clover and lime in connection, has carried up the price of many extensive tracts of land, from $10 to $50 per acre, and has enabled the occupant to raise large crops of wheat, where he could get only small crops of rye; and it has frequently increased his crop of wheat three-fold, where it had been previously an object of attention.

It is a common observation of intelligent farmers, that they are never at a loss to renovate such lands as will produce even a moderate crop of clover. Poor clay lands have been brought to a clover-bearing state, by sowing an early and late crop of oats in the same season, and feeding them off on the ground. Poor, sandy soils may be made to sustain clover, with the aid of manure, ashes and gypsum, combined with the free use of the roller. This object is much facilitated by scattering dry straw over the surface, which affords shade, increases the deposit of dew, and prolongs its effects. Whenever the period of clover-producing is attained, the improvement of the soil may be pushed, with a rapidity commensurate with the inclination and means of the owner.

Harvesting Clover Seed.—After taking off one crop, or pasturing the field till June, or to such time as experience shows to be the most proper, the clover should be kept unmolested, to mature a full crop of seed. Early mowing removes the first weeds, and the second growth of the clover is so rapid, as to smother them and prevent their seeding, and the clover is thus saved comparatively clean. It is then mown and raked into very small cocks, and when dried at the top, they are turned completely over without breaking; and as soon as well dried, they may be carried to the threshing floor, and the seeds beaten out with sticks, light flails, or with a threshing machine.

An instrument with closely-set teeth and drawn by a horse, is sometimes used for collecting the clover heads from the standing stalks, from which the seed is afterwards separated. If wanted for use on the farm, these heads are sometimes sown without threshing. The calyx of the clovers is so firmly attached to the seed, as to be removed with difficulty; but if thrown into a heap after threshing, and gently pressed together, a slight fermentation takes place, and the seed is afterwards readily cleaned. A fan or clover machine may be used for cleaning the seed for market. The produce is from three to six bushels per acre, which is worth to the farmer, from $3 to $5 per bushel, of 60 pounds.

SOUTHERN CLOVER (*T. medium*) is a smaller species than the *T. pratense*, and matures ten or fourteen days earlier. The soil best suited to it, is nearly similar to that required by the northern clover; but it succeeds much better on a light, thin soil than the latter, and it should be sown thicker. Strong clay or rich, loamy soils will produce much heavier crops of the larger kind. Experience alone will determine which of these kinds should be adopted, under all the circumstances of soil and fertility, and the uses for which it is designed.

WHITE CREEPING CLOVER (*T. repens*, Fig. 27).—There are

FIG. 27.

several varieties of white clover, all of which are hardy, nutritious and self-propagating. Wherever they have once been, the ground becomes filled with the seed, which spring up whenever an opportunity is afforded them for growth. They are peculiarly partial to clay lands having a rich vegetable mold on the surface ; and the addition of gypsum, will at all times give them great luxuriance. Their dwarf character renders them unfit for the scythe, while the dense-ly-matted mass of sweet, rich food, ever growing and ever abundant, makes them most valuable for pasture herbage.

THE YELLOW CLOVER, HOP TREFOIL OR SHAMROCK (*T. procumbens*), like the white, is of spontaneous growth, very hardy and prolific. It bears a yellow flower and black seeds. It is one of those unostentatious plants, which though never sown and little heeded, help to make up that useful variety, which gives so much value and permanence to our best pasture lands.

Another variety of the yellow clover grows to the height of 24 to 30 inches, in most of the States, and bears a profusion of flowers and seeds. This is a good forage for sheep, and an excellent fertilizer for the land, but is not much relished by cattle or horses.

MANY OTHER OF THE MINUTE CLOVERS AND LEGUMINOSÆ, THE WILD PEA, and other species of this family, abound in our untilled lands, and add greatly to the nutritive character of the forage, although their merits and even their existence are scarcely known.

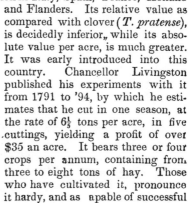

CRIMSON OR SCARLET CLOVER (*T. incarnatum*, Fig. 28) is a native of Italy, and much cultivated in France. It bears a long head, of bright scarlet flowers, and in southern Europe is a profitable crop. Although it was introduced into this country many years since, it has not hitherto commended itself to particular attention as an object of agriculture.

LUCERN (*Medicago sativa*, Fig. 29) is one of the most productive plants for forage, ever grown. It was extensively cultivated by the Greeks, and other nations of antiquity; and it has been a prominent object of attention in Italy, Spain, France, Holland and Flanders. Its relative value as compared with clover (*T. pratense*), is decidedly inferior, while its absolute value per acre, is much greater. It was early introduced into this country. Chancellor Livingston published his experiments with it from 1791 to '94, by which he estimates that he cut in one season, at the rate of $6\frac{1}{5}$ tons per acre, in five cuttings, yielding a profit of over $35 an acre. It bears three or four crops per annum, containing from three to eight tons of hay. Those who have cultivated it, pronounce it hardy, and as capable of successful growth in this country as clover; but to reach the highest product, it requires a richness of soil and carefulness of cultivation, which would give an enormous produce to its more humble rival.

Manner of Cultivation.—It must have a deep, dry, loamy soil, free from weeds, and well filled with manure. A suitable crop to precede it is corn or potatoes, heavily manured and kept clean. Then plow in the fall, and add 40 bushels crushed bones per acre; and early in April, harrow thoroughly, and sow in drills, from one to two and a half feet apart, at the rate of eight to ten pounds of seed per acre. Stir the ground and extirpate the weeds with the cultivator and hoe, carefully pulling out by hand any that may be found in the drills. It may be lightly cropped the first year, and more freely the second, but it does not attain full maturity till the third. The roots strike deep into the ground, and being a perennial, it requires no renewal except from the loss of the plants by casualties. It should be cut before growing too large, and cured like clover.

Liquid manure is good for it, as are also gypsum and ashes. Barn-yard manure is occasionally necessary; but to avoid weeds, it must be thoroughly fermented to destroy the seeds. It is sometimes sown broadcast, but the rapid progress of weeds and grass in the soil, will soon extirpate it if they are suffered to grow; and there is no means of effectually eradicating them but by cultivating the lucern in drills, and the hoe and cultivator can then keep the weeds in subjection. It is one of the most desirable plants for soiling. From the care and attention required, the cultivation of lucern is properly limited to an advanced state of agriculture and a dense population, where labor is cheap and products high. In the neighborhood of large cities, it may be advantageously grown, and in all places where soiling is practiced.

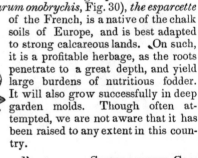

SAN-FOIN (*Hedysarum onobrychis*, Fig. 30), *the esparcette* of the French, is a native of the chalk soils of Europe, and is best adapted to strong calcareous lands. On such, it is a profitable herbage, as the roots penetrate to a great depth, and yield large burdens of nutritious fodder. It will also grow successfully in deep garden molds. Though often attempted, we are not aware that it has been raised to any extent in this country.

BOKHARA OR SWEET-SCENTED CLOVER (*Melilotus major*) is a tall, shrublike plant growing to the height of four to six feet with branches, whose extremities bear numerous small white

FIG. 30.

flowers of great fragrance. When full grown, it is too coarse for forage, but if thick and cut young, it yields a profusion of green or winter fodder. It should be sown in the spring, with about two pounds of seed per acre, in drills 16 to 20 inches apart. It must be kept clear of weeds and cultivated like lucern. It requires a rich, mellow, loamy soil

SPURRY (*Spergula arvensis*, Fig. 31) might probably be  introduced into American husbandry for forage, with decided advantage. It is a hardy plant, and grows spontaneously in the middle States. Its chief merit consists, in its growing on soils too thin to bear clover. On such, it can be judiciously used to bring them up to the clover-bearing point, from which they can be taken, and carried forward much more rapidly by the clovers. Van Voght says, " it is better than red or white clover; the cows give more and bet-

FIG. 31.

ter milk when fed on it, and it improves the land in an extraordinary degree. If the land is to lie several years in pasture, white clover must be sown with it. When sown in the middle of April, it is ripe for pasture by the end of May. . If eaten off in June, the land is turned flat and another crop is sown, which affords fine pasture in August and September. This operation is equivalent to a dressing of ten loads of manure per acre. The blessing of spurry, *the clover of sandy lands*, is incredible when rightly employed." Three crops can be grown upon land in one season, which, if turned in or fed on the ground, can be made a means of rapid improvement to the soil.

PASTURES.

It is too often the case, that pastures are neglected, and like woodlands, are allowed to run to such vegetation as unassisted nature may dictate. As a necessary consequence, their forage is frequently meagre and coarse, and incapable, either in quantity or quality, of supporting half the number of cattle in a poor condition, that might otherwise be fullfed from the same surface. But if we consider, that pastures furnish most of the domestic stock with their only food, for seven months of the year at the North, and generally for nine or ten months at the South, they may well be deemed worthy the particular attention of the farmer.

Pastures ought to be properly divided, and it is a difficult point to determine between the comparative advantages of small ranges, and the expense and inconvenience of keeping up numerous divisions. The latter require a large out lay on every farm, not only for the first cost of materials and the annual repairs, but from the loss of land occupied by them ; and they are further objectionable from their harboring weeds and vermin. Yet it is beneficial to give animals a change of feed; and the grass comes up evenly and grows undisturbed, if the cattle be removed for a while. There is a further advantage, in being able to favor some particular individuals or classes of animals. Thus, fattening stock ought to have the best feed ; milch cows and working animals the next ; then young stock ; while sheep will thrive on shorter feed than either, and greedily consume most plants which the others reject. By this means, a field will be thoroughly cleansed of all plants which animals will eat, and the remainder should be extirpated. The same care ought to be taken to prevent the propagation of weeds in pastures as in other fields. Many of these, mullen, thistles and the like, multiply prodigiously from sufferance, and if unchecked, will soon overspread the farm.

Every pasture, if possible, should be provided with running water and shade trees, or other ample protection against a summer's sun. The last can at all times be secured by a few boards, supported on a light, temporary frame. Excessive heat exhausts, and sometimes sickens animals ; consequently, it materially diminishes the effects of food in promoting the secretion of milk, and the growth of wool and flesh.

All grounds immediately after long rains, whether in early spring or late autumn, and especially, after the winter's frosts have just left them, are much injured by the poaching of cattle, if allowed to run upon them. · Clay lands and those which have been recently seeded, are peculiarly susceptible of injury from this cause ; and from such fields and at such times, every animal should be rigidly excluded. *On late and off early*, is a good rule to be adopted for spring and fall pasturing.

Wherever the grasses disappear, fresh seeds should be added and harrowed in ; mosses must be destroyed; they should be properly drained, and every attention paid to them that is bestowed on the mowing lands, except that they seldom require manures. But ashes, gypsum, lime and bone dust may sometimes be applied to them with great profit. Pastures should take their course in rotation, when they get bare of good herbage or full of weeds, and it is possible to break them up advantageously. Though many choice, natural forage plants may thus be destroyed, yet if again turned into grass at the proper period, and they are sown with a plentiful stock of assorted grass seeds, on a rich and well-prepared surface, they will soon place themselves in a productive state.

CHAPTER VII.

GRAIN AND ITS CULTIVATION.

WHEAT (Triticum).

Fig. 32.   Fig. 33.   Fig. 34.
SPRING BALD WHEAT. WINTER BALD WHEAT WINTER BEARDED WHEAT.

THIS is one of the most important and most generally cultivated of the cereal grains (or grasses as they are bo-tanically termed), though both rice and maize or Indian corn, contribute to the support of a larger population. It is found in every latitude, excepting those which approach too nearly to the poles or equator; but it can be profitably raised,

only within such as are strictly denominated temperate
Linnæus describes but six varieties, yet later botanists enu
merate about thirty, while of the sub-varieties, there are
several hundred.

The only division necessary for our present purpose, is of
the winter wheat (*Triticum hybernum*), and spring or
summer wheat (*T. æstivum*). The former requires the action of frost to bring it to full maturity, and is sown in autumn.   Germination before exposure to frost, does not, however, seem absolutely essential to its success, as fine crops
have been raised from seed sown early in the spring, after
having been saturated with water and frozen for some weeks.
It has also been successfully raised, when sowed early in the
season, while the frost yet occupied the ground.

Spring and winter wheat may be changed from one to the
other, by sowing at the proper time through successive seasons, and without material injury to their character.   The
latter grain is by far the most productive ; the straw is stouter ;
the head more erect and full ; the grain plumper and heavier, and the price it bears in market, from eight to fifteen per
cent. higher than that of spring wheat.   This difference of
price depends rather on the appearance of the flour and its
greater whiteness, than on any intrinsic deficiency in its
substantial qualities.   The analysis of Davy gave in 100
parts of

	Gluten.	Starch,	Insoluble matter.
Spring wheat of 1804,	24	70	6
Best Sicilian winter wheat,	21	74	5
Good English winter wheat of 1803,	19	77	4
Blighted wheat of 1804,	13	53	34

The above analysis gives the greatest nutritive value to
the spring wheat, as the gluten (animalized matter) constitutes the most important element in flour.   It will also be
noticed, that the Sicilian yields about two per cent. more
gluten than the English, which enables the flour to absorb
and retain a much larger proportion of water when made
into bread.   This is what is termed by the bakers, *strength ;*
and when gluten is present in large proportions, other qualities being equal, it adds materially to the value of flour.
American wheat also contains more gluten than English,
and that from the southern States, still more than that from
the northern.   An eminent baker of London says, American
flour will absorb from eight to fourteen per cent. more of its

own weight of water, when manufactured into bread or bis-
cuit, than their own ; and another reliable authority asserts,
that while 14 lbs. of American flour will make 21½ lbs of
bread, the same quantity of English flour will make only
18½ lbs.

As a general rule, the drier or hotter the climate in which
the grain is raised, the greater is the evaporation, and the
more condensed is the farina of the grain, and consequently
the more moisture it is capable of absorbing when again exposed to it. Certain varieties of wheat possess this quality
in a higher degree than others. Some manures and some
soils also give a difference with the same seed ; but for ordinary consumption, the market value (which is the great
consideration with the farmer), is highest for such wheat as
gives the largest quantity of bright flour, with a due proportion of gluten. Other prominent differences exist among
the leading cultivated varieties of wheat, such as the bearded
and bald or beardless ; the white and red chaff; those having large and strong stalks ; or a greater or a less tendency
to tiller or to send out new shoots. There is great room for
selection in the several varieties, to adapt them to the different soils, situations, and climate for which they are designed.

Preparation of the land for sowing. Wheat is partial to
a well-prepared clay or heavy loam, and this is improved,
when it contains either naturally or artificially, a large proportion of lime. Many light, and all marly or calcareous
soils, if in proper condition, will give a good yield of wheat.
Lime is an important aid to the full and certain growth of
wheat, checking its exuberance of straw and liability to
rust, and steadily aiding to fill out the grain. A rich, mellow turf or clover ley is a good bed for it ; or land which
has been well manured and cleanly cultivated in roots or
corn the preceding year.

Fresh barn-yard manure applied directly to the wheat crop,
is objectionable, not only from its containing many foreign
seeds, but from its tendency to excite a rapid growth of weak
straw, thus causing the grain both to lodge and rust. The
same objection lies against sowing it on rich, alluvial or vegetable soils ; and in each, the addition of lime or ashes, or both,
will correct these evils. A dressing of charcoal, has in
many instances been found an adequate preventive ; and so
beneficial has it proved in France, that it has been extensively introduced there for the wheat crop. A successful example of uninterrupted cropping with wheat, through several
6*

years, has been furnished by a Maryland farmer, who used fresh barn-yard manure, with lime. But this is an exception, not a rule; and it will be found that profitable cultivation requires, that wheat should take its place in a judicious rotation. The great proportion of silica in the straw of cereal grains, (amounting in wheat, barley, oats and rye, to about four fifths of the total of ash from the grain and straw), shows the necessity of having ample provision made for it in the soil, and in a form susceptible of ready assimilation by the plant. This is afforded by ashes, and from the action of lime upon the soil.

Depth of Soil is indispensable to large Crops.—The wheat plant has two sets of roots, the first springing from the seed and penetrating downwards, while the second push themselves laterally, near the surface of the ground, from the first joint. They are thus enabled to extract their food from every part of the soil, and the product will be found to be in the ratio of its extent and fertility. Under-draining and sub-soil plowing contribute greatly to the increase of crops, and it is essential that all surface water be entirely removed Wheat, on heavy clay lands, is peculiarly liable to winter kill, unless they are well-drained. This is owing to successive freezing and thawing, by which the roots are broken or thrown out. When this is done to a degree that will materially diminish the crop, the naked spots may be sown with spring wheat. Any considerable portion of the latter, will lessen the value for sale, but it is equally good for domestic use. The land should be duly prepared for the reception of the seed, by early and thorough plowing, and harrowing, if necessary.

Selection and preparation of Seed.—Many persons select their seed by *casting*, or throwing the grain to some distance on the floor, using only such as reaches the farthest. This is a summary way of selecting the heaviest, plumpest grain, which if Sprengel's theory be correct, is attended with no advantage, beyond that of separating it from the lighter seeds of chess or weeds. It is certain, that the utmost care should be taken in removing everything from it but pure wheat, and this should be exclusively of the kind required. When wheat is not thoroughly cleaned by casting, a sieve or riddle may be used; or it should even be picked over by hand, rather than sow anything but the pure seed.

Previous to sowing, a strong brine ought to be made of salt and water, and in this the grain is to be washed for five minutes, taking care to skim off all light and foreign seeds. If the grain be smutty, this washing should be repeated in another clean brine, when it may be taken out and intimately mixed with one twelfth its bulk of fresh pulverized quick-lime. This kills all smut, cleans out weeds from the grain, and insures early and rapid growth When the seed is not smutty, it may be prepared by soaking or sprinkling with stale urine, and afterwards mixed with the lime; and if well done, this also will prevent smut, though the first is most certain. (See varieties of seed following, for further directions.)

Quantity of Seed and time of sowing.—On well pulverized, ordinary wheat soils, about five pecks of seed are sown to the acre, while rough land, clay soils and such as are very fertile, require from six to eight. In Maryland, but three pecks are frequently sown, and some of the best crops have been raised from only two pecks of seed to the acre, on a finely-pulverized soil. It takes more seed when full and plump than when shrunken, as there may be nearly two of the latter to one of the former, in the same measure. A difference is to be observed according to the wheat, some needing more than others. A large quantity of seed, produces an earlier growth of light straw and head, but does not usually increase the aggregate crop. There is always a tendency in wheat and most of the cereal grasses, to tiller or send out new shoots for future stalks. This is a law of these plants, which compels them to make the greatest effort to cover the whole ground; and sometimes a single seed will throw out more than 100 stalks. In early sowing, the wheat tillers in the autumn; in late sowing this is done in part only, till the ensuing spring. Thick sowing, is a substitute for tillering, to the extent that would otherwise be induced, and is equivalent to an earlier sowing of a smaller quantity. The time for sowing in the northern States, is from the 10th to 20th September. If sown earlier, it is liable to attack from the Hessian fly, and if later, it does not have time to root as well; and is in more danger of being thrown out by the frosts or of winter killing. Late sowing is also more subject to rust the following season, from its later ripening.

Sowing.—When the ground has been well mellowed, the seed may be sown broadcast and thoroughly harrowed in Rolling is a good practice, as it presses the earth closely up on the seed and facilitates germination; and as soon as the seed is covered, the water furrows should be cleaned out.

and again late in autumn, and early in the following spring
In northern Europe, it has been found a preventive against
winter killing on strong clays, to sow the wheat in the bottom of each furrow, six inches deep, and cover it with the
succeeding one.   The wheat thus planted, comes up as soon
as on the fields sown broadcast and harrowed, grows more
vigorously, withstands the winters and produces large crops.
Plowing in wheat with a light furrow, is perhaps, under any
circumstances, better than harrowing, as the wheat is thereby all buried, and at a more suitable depth than can be done
by the harrow.   The roughness of the furrows when left
without harrowing, is advantageous in heavy or clay lands,
and only injurious in light or sandy.

After Culture.—Harrowing in the spring, adds to the
growth of the crop, by loosening the soil; and the loss of
the few plants thus destroyed, is much more than compensa-
ted by the rapid tillering and vigor of those which remain.
Sowing in drills and hoeing between them, is much practiced in Europe.   The additional amount thus frequently
raised, would seem to justify the adoption of this mode of
cultivation in this country; and it should at least be done,
so far as to give it a fair trial.   On light soils, rolling the
wheat both in fall and spring, is highly advantageous.
When the growth is luxuriant, decided benefit has attended
feeding off the wheat on the field in the fall or spring, taking care to permit the animals to go on, only when the
ground is firm.

Enemies of Wheat.—These are numerous.   It is subject
to the attack of the Hessian fly, if sown too early in the
fall, and again the ensuing spring, there being two annual
swarms of the fly, early in May and September.   When
thus invaded, harrowing or rolling, by which the maggots
or flies are displaced or driven off, is the only remedy of
much avail.   Occasionally, other flies, and sometimes wheat
worms commit great depredation.   There is no effectual
remedy known against any of these marauders, beyond rolling, brushing and harrowing.   Dusting the grain with lime,
ashes and soot, have been frequently tried, as have also the
sprinkling them with urine, dilute acids and other liquids or
steeps.   Fumigating them in the evening, when the smoke
creeps along through the standing grain, has been often tried,
but without decided success.   For this last purpose, a smoul-
dering heap of damp brush, weeds or chips, is placed on the
windward side of the field; and its efficacy may be increased

by the addition of brimstone. Whenever obnoxious to these attacks, the only safety is, to place the crop in the best condition to withstand them, by hastening its growth, and by the propagation of the most hardy varieties. An application of unleached ashes in damp weather, will sometimes diminish the ravages of worms at the root. Quick-lime has the same effect on all insects with which it comes in contact; but it should be carefully applied to avoid injury to the plants.

Smut is a dark brown or blackish, parasitic fungus, which grows upon the head and destroys the grain. The only remedy for this, is washing the seed in two or three successive strong brines, and intimately mixing and coating it with quick-lime before sowing.

Rust affects the straw of wheat while the grain is forming, and before it is fully matured. It is almost always present in the field, but is not extensively injurious except in close, showery and hot (muggy) weather. The straw then bursts from the exuberance of the sap, which is seen to exude, and a crust or iron-colored rust is formed in longitudinal ridges on the stalk. It is generally conceded, that this rust is a fungus or minute parasitic plant which subsists on the sap; but whether it be the cause or consequence of this exudation is not fully determined. There is no remedy for this when it appears, and the only mitigation of its effects, is to cut and harvest the grain at once. The straw in this case will be saved, and frequently, a tolerable crop of grain, which partially matures after cutting; while if suffered to stand, both straw and grain will be almost totally lost. The only preventives experience has hitherto found, are the selection of hardy varieties of grain, which partially resist the effects of rust; sowing on elevated lands where the air has a free circulation; the abundant use of saline manures, salt, lime, gypsum, and charcoal; the absence of recent animal manures; and early sowing, which matures the plant before the disease commences its attack.

Harvesting.—The grain should be cut immediately after the lowest part of the stalk becomes yellow, while the grain is yet in the dough state, and easily compressible between the thumb and finger. Repeated experiments have demonstrated, that wheat cut at this time, will yield more in measure, of heavier weight, and a larger quantity of sweet, white flour. If early cut, a longer time is required for curing before storing or threshing.

Threshing is usually done among extens.ve farmers, with some one of the large machines taken into the field, and driven by horse power. The use of these enables the farmer to raise some of the choicest kinds of grain, whose propagation before their introduction was limited from the great difficulty of separating the grain from the head. He can also push his wheat into market at once, if the price is high, which is frequently the case immediately after harvest ; and he saves all expense and trouble of moving, storing, loss from shelling and vermin, interest and insurance. For the moderate farmer, a small single or double horse machine, or hand threshing in winter, where there is leisure for it, is more economical than the six or eight horse-thresher.

Mowing or Stacking.—When stored in the straw, the grain should be so placed as to prevent heating or moulding.

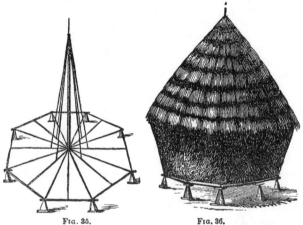

Fig. 35.   Fig. 36.

Unless very dry, when carried into the barn, this can only be avoided by laying it on scaffolds, where there is a free circulation of air around and partially through it. If placed in a stack, it should be well elevated from the ground ; and if the stack be large, a chimney of lattice or open work should be left from the bottom, extending through the centre to the top ; or a large bundle may be kept at the surface in the middle, and drawn upwards as the stack rises, thus leaving an opening for circulation, entirely through the centre of the grain. Additional security would be afforded by similar openings at suitable intervals, in a horizontal direction. Mice and rats may be avoided, by laying the foundation of the stack on posts or stones, elevated beyond their reach, and covered at the top with projecting caps. Weevils sometimes affect the grain after storing. These may be almost if not wholly prevented, by thorough cleanliness of the premises where the grain is stored.

The cut on the preceding page, Fig. 35, shows a frame for stacking, combining the advantages of circulation through the centre, and an elevation which secures the grain from the depredations of vermin. Fig. 36, shows a stack complete, which is better made and more neatly finished than is too often done in this country. It is an important item of husbandry, so to stack grain as to avoid loss from the admission of rain. No inconsiderable share of the stack, is thus frequently destroyed.

The straw and chaff of wheat should never be wasted. This is the most nutritious of the cereal straws. It yields good fodder to cattle in time of scarcity, and is always good for this object, when cut and mixed with meal or roots; and particularly, when early harvested and well cured. Turneps and straw are the only food of half the cattle, and most of the sheep, throughout Great Britain, and no where do they thrive more rapidly, or better remunerate their owners, than in that country. It is of great use also, as bedding for cattle and as an absorbent of animal and liquid manures. It furnishes in itself the best manure for succeding grain crops; containing large proportions of the salts or ash required. When threshed on the field and not wanted for cattle, it should be scattered over the ground, and either plowed in or suffered to decay on the surface.

VARIETIES OF SEED.

Much depends on the judicious selection of seed. Some soils are peculiarly adapted to wheat growing, and on these should be sown the finest varieties, which are generally of a more delicate character. Wheat on other soils is liable to many casualties, and on such, only the hardier kinds should be propagated. Careful and repeated trials with different varieties of seeds, on each field or on those which are similar, will alone determine their adaptation to the soil. There are several choice varieties of winter wheat in cultivation in the United States, some of which stand higher in one, and some in another section. Some in high repute abroad, have been introduced into this country, and proved to be valuable acquisitions; while others have been found decidedly inferior to many of the long adopted varieties. Experiment alone will enable the farmer to decide as to their value for his own grounds, however high they may stand elsewhere. When of a fine quality, and found to produce well on any given soils, their place should not be usurped by others, till repeated trials have shown their superiority, either in yield or character. But when the acclimated grain is inferior, other seed from remote distances, even if no better in quality, may properly be substituted for it, as a decided benefit has been found to follow a change.

Wheat and nearly all seeds are found to be more productive, when taken from a soil inferior to the one intended for sowing; and it is claimed that such as have been produced, either in a warmer or colder climate, will mature earlier. It is not essential that the fullest, heaviest grain be sown. Sprengel affirms, that seed somewhat shrunken, is more certain to give a good yield than the choicest seed; and numerous trials would seem to favor this conclusion. The grain designed for seed should be well ripened before harvesting. From the ever-varying character of the different kinds of seed, their superiority at one time and on one locality, and their inferiority at other times and in other situations, it seems almost superfluous to give a particular enumeration of the present most popular kinds. A brief mention of such only, as stand high in public favor in this country, with some of their most striking peculiarities, is all that our limits will admit.

The Improved Flint is extensively cultivated in the fine wheat-growing country of western New York, where it was introduced in 1822. It is hardy, and withstands the winters remarkably well. A striking improvement in the strength of its straw has been observed, which at first inclined to lodge, but it is now erect and firm till fully ripened. The heads are also fuller and longer than when first introduced; the berry is plump and white, yielding a large proportion of choice flour; and it is retained in the head with greater tenacity, which is a decided advantage in harvesting, where threshing machines are substituted for the flail.

The Old Genesee Red Chaff is a bald, white wheat, first cultivated in the same region, in 1798; and for a long time it was the decided favorite. Since 1820, however, it has been very subject to rust and blast; but when circumstances are favorable, it is still found to be highly productive. In other localities, its cultivation may be attended with the most satisfactory success.

The White May of Virginia was a choice variety, and extensively raised in the neighborhood of the Chesapeake Bay, in 1800, but is now nearly extinct there. It has been cultivated in New York for ten years, is a good bearer and very heavy, weighing frequently 66 lbs. per bushel. It ripens early, in consequence of which, it escapes rust.

The Wheatland Red is a new variety, discovered and propagated by Gen. Harmon, of Monroe Co., N. Y., by whom it is held in high estimation. It produces well and ripens early.

The Kentucky White-Bearded, Hutchinson or *Canadian Flint* is very popular in western New York, where it has been rapidly disseminated since its first introduction, some twelve or fifteen years since. It is hardy, a good yielder, with a short, plump berry, weighing 64 lbs. per bushel. It requires thicker sowing (about 25 per cent. more seed) than the improved flint, as it does not tiller as well, and unlike that, it shells easily, wasting much unless cut quite early.

The English Velvet Beard or *Crate Wheat* has a coarse straw, large heads, a good berry of a reddish hue, and is well suited to the rich, alluvial, bottom lands, where its firm straw prevents its lodging. It is a fair yielder and tolerably hardy; but its long beard is a great objection to its introduction on such lands as are suited to the finer kinds.

The Yorkshire or *English Flint* or *Soules Wheat* has been recently introduced, and is similar in its leading features to the old Genesee.

The White Provence is a new and favorite variety, but its slender stalk frequently subjects it to lodging. It is only suited to the finest calcareous wheat soils.

The Blue Stem has been raised with great success in Pennsylvania, where it resisted smut and rust when all other kinds in the vicinity, were affected by it.

The Mediterranean is a coarse wheat with a thick skin, yielding a dark flour. It resists rust and the fly, is a good bearer, and may be profitably grown where other choice kinds fail.

The Egyptian, Smyrna, Reed, Many-Spiked, or *Wild Goose Wheat* is also a hardy variety, with a thick straw which prevents its lodging.

PRODUCTION OF NEW VARIETIES OF WHEAT.

Besides introducing valuable kinds from abroad, and the improvement of such as we now have by careful cultivation, new varieties may be secured by hybridizing or crossing This is done by impregnating the female organs of the flowers on one plant, by the pollen from the male organ of another. The progeny sometimes differs materially from both parents, and occasionally partakes of the leading qualities of each. Among those thus produced, some may be found of peculiar excellence, and worthy of supplanting others, whose value is declining. The effect of this crossing, is striking in the ear of corn, where the red and white, the blue and yellow kernels are seen to blend in singular confusion over the whole ear, each differing, too, in size, shape and general qualities. Observation will sometimes detect a new variety of wheat in the field, self-hybridized, the result of an accidental cross. If this has superior merit, it should be carefully secured and planted in a bed by itself for future seed.

Propagation may be extended with incredible rapidity by dividing the plant. The English Philosophical Transactions give the result of a trial, made by planting a single grain on the 2d of June. On the 8th of August, it was taken up and separated into 18 parts, and each planted by itself. These were subdivided and planted, between 15th of September and 15th of October, and again the following spring. From this careful attention, in a fertile soil, 500 plants were obtained, some containing 100 stalks bearing heads of a large size; and the total produce within the year, was 386,840 grains from the single one planted.

SPRING WHEAT.

This requires a soil similar to that of winter grain, but it should be of a quick and kindly character, as the grain has a much shorter time to mature. The ground must be well pulverized and fertile. The best crops are raised on land that has been plowed in the fall, and sown without additional plowing, taking care to harrow in thoroughly. When planted early, the wheat rarely suffers from the fly, as it attains a size and vigor that withstands any injury from the fly when it appears. In certain localities, where the fly abounds and the wheat has not been early sown, it is found necessary to keep back the young plants, till the disappearance of the fly. Large crops have been obtained under favorable circumstances, when sown as late as the 20th May

VARIETIES.

The Black Sea Wheat is one of the most popular kinds at present cultivated. Of this there are two varieties, the red and the white chaff, both of which are bearded. The former is generally preferred. This wheat has yielded very profitable crops. The flour from this, like that from the Mediterranean wheat, is of a dark color.

The Siberian is an excellent wheat, and has been much raised in this country. It produces a full, fine grain, is hardy and a good bearer. The *Italian* has also been extensively cultivated, and held in high estimation; but it is now generally giving place to the preceding, where both have been tried.

There are some other varieties which bear well and are tolerably hardy. Excellent spring grain has been produced, by early sowing from choice winter wheat, which has retained most of the characteristics of the original, under its new summer culture. In large sections of this country, wheat has been seriously injured by winter-killing and other casualties; and wherever these prevail, and the soil is suited to it, spring or summer wheat may be advantageously introduced. A proper attention to the selection of seed and the preparation of the soil, will generally insure a profitable return. If the market value of this wheat is not as high as the winter grain, it may at least afford all that the farmer and his laborers require as food; and he will generally find, if not in a wheat-growing region, that he can dispose of his surplus crop among his neighbors before the next harvest comes round, and at satisfactory prices.

RYE (Secale sereale).

This is extensively cultivated in the northeastern and middle Atlantic States. It is grown on the light lands of Ohio and Michigan, and as the supporting elements of wheat become exhausted in the soil of the rich agricultural States of the West, rye will take its place in a great measure on their lighter soils. Most of the eastern States produced wheat when first subjected to cultivation; but where lime did not exist in the soil, the wheat crop soon failed, and it gradually receded from the Atlantic border, except in marly or calcareous soils, rye almost universally succeeding it. But the liberal use of lime, connected with an intelligent application of the agricultural improvements of the present day, are regaining for wheat, much of its ancient territory.

Rye resembles wheat in its bread-making properties, and

for this purpose it is only second to wheat, in those countries where it is cultivated. There is a peculiar aroma connected with the husk of the grain, which is not found in the finely-bolted flour. The grain when ground and unbolted, is much used in the New England States, for mixing into loaves with scalded Indian meal ; it is then baked for a long time, and is known as *rye-and-Indian* or *brown bread*. This possesses a sweetness and flavor peculiar to itself, which is doubtless owing in no small degree, to the quality above mentioned. Von Thaer says " this substance appears to facilitate digestion, and has a singularly strengthening, refreshing and beneficial effect on the animal frame." Rye is more hardy than wheat, and is a substitute for it on those soils which will not grow the latter grain with certainty and profit.

Soil and Cultivation.—Neither strong clay nor calcareous lands are well suited to it. A rich sandy loam is the natural soil for rye, though it grows freely on light sands and gravels, which refuse to produce either wheat, barley, or oats. Loamy soils that are too rich for wheat, and on which it almost invariably lodges, will frequently raise an excellent crop of rye, its stronger stem enabling it to sustain itself under the luxuriant growth.

The preparation of the Soil for Rye, is similar to that for wheat; and it may be advantageously sown upon a rich old turf or clover ley, or after corn or roots where the land has been well manured, and thoroughly cleansed from weeds. There is not an equal necessity for using a brine-steep for rye as for wheat, yet if allowed to remain a few hours in a weak solution of saltpetre or some of the other salts, it promotes speedy germination and subsequent growth.

Cultivation.—There is but one species of rye ; but to this cultivation has given two leading varieties, the spring and winter. Like wheat, they are easily transformed into each other, by sowing the winter continually later through successive generations, to change it into spring rye, and the opposite course will ensure its re-conversion into winter grain. The last should be sown from the 20th of August to the 20th of September, the earliest requiring less seed, as it has a longer time to tiller and fill up the ground. Five pecks is the usual quantity sown, but it varies from one to two bushels according to the quality of the soil, the richest lands demanding most.

It is a practice among many farmers, to sow rye on light lands, among their standing corn, hoeing it in, and leaving the ground as level as possible. On such lands, this is attended with several advantages; as it gives the grain an early start, and a moist, sheltered position, at a time when drought and a hot sun would check or prevent vegetation. As soon as the corn is matured, it is cut up by the roots and placed in compact shocks, or removed to one side of the field, when the rye is thoroughly rolled. When sown on a fresh plowed field, it should be harrowed in before rolling.

Great success has attended the turning in of green crops, and following the fresh plowing with immediate sowing of the seed. This brings it forward at once. No after cultivation is needed, except harrowing in the spring, and again rolling, if the land is light, both of which are beneficial; for though some of the stools may be thus destroyed, the working of the ground assists the remaining plants, so as to leave a great advantage in favor of the practice. A friend of the writer had occasion to plow some land in the spring, which joined a field of rye belonging to a neighbor. The owner claimed damages for supposed injury by the team and plow, which it was agreed should be assessed, on examination after harvesting, when it appeared that the *damaged* part was the best of the whole field. An honest English yeoman received several pounds from a liberal squire, for alleged injury to his young grain, from the trampling of horses and hounds in a fox chase ; but at harvest, he found the crop so much benefitted by the operation, that he voluntarily returned the money. If the rye is luxuriant, it may be fed both in the fall and spring. Early cutting, as in wheat, produces moré weight, larger measure and whiter flour. But whatever is intended for seed, must be allowed to ripen fully on the ground.

Southern Rye differs materially in its manner of growth, from that cultivated in the North. I believe, however, this difference arises exclusively from dissimilarity of climate , and that, like the sectional sub-varieties of corn or maize, a few years' successive growth in a peculiar latitude, will give to either species, the same characteristics as the longer acclimated grain. It tillers remarkably, and grows with great luxuriance during fall and a part of winter, affording excellent forage for cattle, sheep, and other animals When the animals are taken off the following spring, the grain runs up to seed, yielding from 10 to 15 bushels of ripened grain to the acre. I saw a beautiful field of this, late in November, adjoining the mansion of Col. Wade Hampton, of South Carolina, which was devoted to the pets of the stables and yards; and especially to the numerous varieties of fowls (aquatic and others), that seemed to revel on their fresh green pastures, in the absence of other herbage.

Diseases.—Rye is subject to fewer casualties than wheat. *Ergot or cockspur* frequently affects it. This fungus is discovered, not only on rye, but on other plants of the order *graminæ.* Several of these elongated, curved and brownish spurs appear on a single head, and they are most frequent in hot, wet seasons. They are poisonous to both man and beast; and when eaten freely, they have generated fatal epidemics in the community, and emaciation, debility, and in some cases death, to animals consuming it. The sloughing of the hoofs and horns of cattle, has been attributed to ergot in their grass and grain. *Rust* like that which affects the wheat crop, and owing probably to the same causes, attacks rye. When this happens, it should be cut and harvested without delay.

Rye for Soiling is sometimes sown by those who wish forage late in autumn and early in spring. For this purpose, it should be sown at the rate of three or four bushels per acre. If on a fertile soil and not too closely pastured, it will bear a good crop of grain; and in some cases when too rank, early feeding will strengthen the stalk and increase the grain

BARLEY (Hordeum, Fig. 33).

Barley is a grain of extensive cultivation and great value. Like wheat and rye, it is both a winter and spring grain, though in this country, it is almost universally sown in the spring. There are six varieties, differing in no essential points, and all originating from the same source. Loudon says, in choosing for seed, "the best is that which is free from blackness at the tail, and is of a pale, lively yellow, intermixed with a bright, whitish cast; and if the rind be a little shrivelled, so much the better, as it indicates thin skin. The husk of thick-rinded barley is too stiff to shrink, and will lie smooth and hollow, even when the flour is shrunk within. The necessity of a change of seed from time to time, for that grown in a different soil, is in no instance more evident than in this grain, which otherwise becomes coarser every successive year. But in this, as in all other

Fig. 33.   grain, the utmost care should be taken that the seed is full bodied."

The principal varieties are the two and six rowed; the last being preferred for hardiness and productiveness in Europe. The first is generally cultivated in this country, from its superior fullness and freedom from smut. There are numerous sub-varieties, such as the *Hudson's Bay*, which ripens very early and bears abundantly ; the *Chevalier* and *Providence*, both accidental, of which a single stalk was first discovered among others of the ordinary kinds, and proving superior and of luxuriant growth, they were widely propagated ; the *Peruvian, Egyptian,* and others. New varieties may be produced by crossing, as with wheat.

Soil.—Barley requires a lighter soil than will grow good wheat, and a heavier than will bear tolerable rye ; but in all cases it must be one that is well drained. A mellow rich loam, ranging between light sand or gravel, and heavy clay is best suited to it.

Cultivation.—It may be sown as soon as the ground is sufficiently dry in spring, on a grass or a clover ley turned over the preceding fall ; or it may follow a well-manured and cleanly-hoed crop. If sown on a sod, it should be lightly plowed in, but not so deep as to disturb the sod, and afterwards harrowed or rolled. The soil must always be well pulverized. From $1\frac{1}{2}$ to $2\frac{1}{2}$ bushels per acre is the usual allowance of seed, poor and mellow soils and early sown, requiring the least. Barley ought never to follow the other white grains, nor should they succeed each other, unless upon very rich soil. No farmer can long depart from this rule, without serious detriment to his soil and crops. Barnyard manures must never be applied directly to this grain, unless it be a light dressing of compost on indifferent soils ; or in moderate quantity after the plants have commenced growing in spring. When the plants are four or five inches high, rolling will be of service if the ground is dry and not compact. This operation gives support to the roots, destroys insects multiplies seed-stalks, and increases their vigor.

Destroying Weeds in Grain.—When grain is infested with cockle, wild mustard or other weeds, they should be extirpated by hand before they are fairly in blossom. If neglected till sometime after this, the seed is so well matured as to ripen after pulling, and if then thrown upon the ground, they will defeat the effort for their removal. When too luxuriant, barley like rye, may be fed off for a few days, but not too closely.

The Harvesting of barley must be seasonably done.

More caution is requisite in cutting it at the proper time, than is necessary to observe with any other grain ; for if cut too late, its extreme liability to shell will cause much waste, and it will shrivel, if cut before it is fully matured. It may be stacked like wheat.

The uses of Barley are various and important. In Europe, it forms no inconsiderable part of the food of the inhabitants. The grain yields from 80 to 86 per cent. of flour, which, however, contains but six per cent. of gluten; seven per cent. being saccharine matter, and 79, mucilage or starch. It is inferior in nutriment to wheat and rye, but superior to oats. In this country, it is principally used for malting and brewing, and in some cases for distilling. When ground, it is more generally appropriated to fattening swine, though sometimes used for other stock.

THE OAT (Avena sativa, Fig. 34).

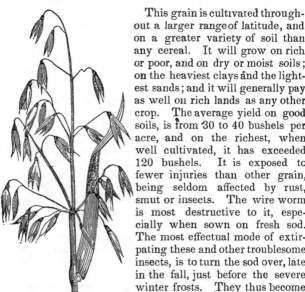

This grain is cultivated throughout a larger range of latitude, and on a greater variety of soil than any cereal. It will grow on rich or poor, and on dry or moist soils; on the heaviest clays and the lightest sands ; and it will generally pay as well on rich lands as any other crop. The average yield on good soils, is from 30 to 40 bushels per acre, and on the richest, when well cultivated, it has exceeded 120 bushels. It is exposed to fewer injuries than other grain, being seldom affected by rust, smut or insects. The wire worm is most destructive to it, especially when sown on fresh sod. The most effectual mode of extirpating these and other troublesome insects, is to turn the sod over, late in the fall, just before the severe winter frosts. They thus become chilled, and incapable of seeking a safe retreat from their fatal effects. If not plowed at that time, it should be done immediately before sowing in spring, when by turning them into the bottom of the furrow, they

Fig. 34.

cannot find their way to the surface in time to injure the plant seriously, before it gets beyond the reach of their attacks.

Varieties.—Of these, Loudon mentions nine as being well defined and entirely distinct, besides which there are many local or recent sub-varieties. He says,

" *The White* or *Common oat* is in most general cultivation in England and Scotland, and is known by its white husk and kernel.

The Black oat, known by its black husk and cultivated on poor soils in the north of England and Scotland.

The Red oat, known by its brownish red husk, thinner and more flexible stem and firmly-attached grains. It is early, suffers little from winds, meals well, and suits windy situations, and a late climate.

The Poland oat, known by its thick white husk, awnless chaff, solitary grains, short white kernel, and short stiff straw. It requires a dry warm soil, but is very prolific.

The Black Poland oat is one of the best varieties; it sometimes weighs 50 lbs. to the bushel.

The Friezland or Dutch oat has plump thin-skinned white grains mostly double, and the large ones sometimes awned. It has longer straw than the Poland, but in other respects resembles it.

The Potato oat has large, plump, rather thick-skinned, white grains, double and treble, with longer straw than either of the two last. It is now almost the only kind raised in the north of England and south of Scotland, and brings a higher price in London than any other variety.

The Georgian oat is a large-grained, remarkably profitable variety and on rich soil, in good tilth, has produced more than any other variety.

The Siberian or Tartarian, is by some conceded a distinct species. The grains are black or brown, thin and small, and turned mostly to one side of the panicle, and the straw is coarse and reedy. It is little cultivated in England, but is found very suitable for poor soils and exposed situations.

The Winter oat is sown at the rate of two bushels per acre in October, the plants are luxuriant and tiller well, and afford good winter and spring pasture for ewes and lambs, and when these are shut out, it affords an ample crop of grain in August."

The Imperial oat is the heaviest raised in the United States, and by many is preferred to all others. It is a clean, bright, plump, heavy grain, yielding a large proportion of flour and nutritive matter. It is hardy and prolific in the northern and middle States.

The Hopetown oat originated from a single stalk, first discovered in 1824, by Mr. Sheriff, in a field of potato oats It is distinguished by its exceeding height, and superior pro duce when sown on rich soils.

The Dyock oat is a recent sub-variety of the Potato oat, and it is claimed for it, that it exceeds the last in the number of bushels yielded per acre, and also in the weight of the grain and the quantity of meal.

The Skinless oats, greatly commended in Ireland, have been tried in this country without much success. They have shown a tendency to degenerate rapidly, the necessary effect of previous highly-artificial cultivation.

There are many other varieties which have a partial or local popularity. From the readiness with which new kinds are produced, careful attention and observation on the part of the farmer, will enable him to detect from time to time, such as may have a decided value over others for particular localities. A superior kind was discovered in a field of common oats in Oneida County, N. Y., some years since, and from the produce of one stool, it became widely disseminated, and has uniformly proved both hardy and prolific.

The variety most cultivated in the United States is the *Common White oat*, which is hardy and a good bearer, weighing from 30 to 34 lbs. per bushel. The *Black oat* is preferred in western New York, and some other sections of the country. Repeated trials have been made with the *Potato oat*, a heavy grain, weighing from 35 to 45 lbs per bushel, but its merits have not hitherto proved conspicuous, enough to justify its usurping the place of the older and long-tried varieties.

The Egyptian oat is much cultivated south of Tennessee, and is said to be the only oat that will mature with certainty in the southern States. It is a grain of medium size, but plump and heavy ; sound, hardy and moderately prolific. It is sown in autumn, and after yielding winter and spring pasturage, gives from 10 to 20 bushels of ripened grain to an acre.

Cultivation.—In this country, oats are sown at the rate of two to four bushels per acre, during all the spring months, and sometimes, though rarely, in June. The earliest sown are usually the heaviest and most productive. They may occupy a turf, or follow any of the well-manured, hoed crops, as mentioned in the preceding grains. No apparent advantage has been derived from steeps for the prevention of smut as in wheat, the impervious husk of the oat, apparently arresting the liquid, and preventing its penetration to the kernel. Sowing salt broadcast over the land, at the rate of two to six bushels per acre, has been found of use to the crop, both in furnishing it with a necessary manure and by killing insects. The seed should be well harrowed in and rolled, and no after attention is required, except to destroy the prominent weeds.

Harvesting.—Oats frequently ripen unevenly, and if there is a large proportion of such as are backward, the proper time for cutting will be, as soon as the grain in the latest, may be rubbed out of the straw by hand. The oat is sufficiently matured for harvesting after it has passed the milk state, and is easily compressed between the thumb and finger. The lower part of the stalk will then have assumed a yellow color, and ceases to draw nutriment from the soil. If cut at this time, the straw is better for fodder and other uses, the grain is fuller; the husk lighter; and the loss from shelling, which is frequently a great item when left too late, is avoided. Oats, when very tall, are most profitably cut with the sickle, and when lodged, with the scythe; but when erect and of medium height, with the cradle, or an approved reaping machine, which is by far the most speedy and economical; and this leaves them in a suitable position for binding into sheaves. They may be stacked like wheat.

The uses of oats are various, and differ materially in different countries. In Scotland, Ireland and many other countries, oat meal is much used as human food; and for this, the Imperial oat or some one of the heavy kinds is preferred, as they afford a larger proportion of meal and less of husk. Scotland draws no inconsiderable part of the support for her entire laboring population, from this meal. It is formed into small thin cakes and eaten with milk, butter or molasses, or it is mixed with water or milk and made into a kind of pudding, under the name of *stirabout,* a favorite dish, which is said to be palatable to those accustomed to it.

They are but little used for human food in this country, and only by emigrants, who bring their early habits with them. They are prepared by kiln-drying and hulling, then grinding and bolting, when required to separate the flour. The meal is scalded before using, and mixed with about half its weight of wheat flour, when made into bread. It is sold by the apothecaries to invalids, for whom it is valuable, from its light, digestible character. It is also stirred into water, making an excellent beverage for laborers in hot weather. The principal use of oats in the United States, is as food for working animals, for which they are unrivalled. Oats are sometimes used when ground for fattening cattle, sheep and swine; but for this purpose, they are surpassed by corn, barley, peas or boiled potatoes. They are an excellent fodder for stock sheep, and for them, are most economically fed in the straw. All stock will do well upon them, when harvested early, and cut previous to feeding, in a suitable cutting box.

Analysis.—Davy found in 1,000 parts of Scotch oats, 743 of soluble or nutritive matter, containing 641 of mucilage or starch, 15 saccharine matter, and 87 gluten or albumen. Those of England, gave 59 of starch, six of gluten, two of saccharine matter, and 33 of husk in 100 parts.

INDIAN CORN, (Zea maize).

Fig. 37.

This next to the grasses, is by far the most important crop of the United States. The quantity this country is capable of raising, would fail to command belief, even if fairly stated. Its capacity will never be fully known, till a demand from abroad shall stimulate production much beyond what it has ever permanently realized heretofore. The census return for 1840, gave 387,000,000 bushels; and for 1843, the estimate of the whole product of Indian corn in this country, was over 400,000,000 bushels. The effect of this immense production of a staple article, is felt in every department of our agriculture; and is conclusively shown by the low prices of beef, pork, mutton, human food, whiskey and high wines, to all of which, corn is made largely to contribute. Nearly all the beef and pork of the vast and fertile West, and much in the North and South is made from it.

Corn seems to have been created for this western hemisphere. It is raised in boundless luxuriance, from the frozen regions of Canada, almost to the Straits of Magellan. It riots in the fierce blaze of our cloudless western sun, and it is here that it attains the highest perfection. Its most prolific area on this continent, lies between 42° North, and 38° South latitude, deducting a limited portion of the equatorial regions. Close attention in its cultivation is necessary when receding from these limits towards the poles, on account of a deficiency of sun for ripening it. In such localities, the smaller and earlier kinds should be planted on a warm soil, so as to mature before the first frosts.

Varieties.—There is no one of the cereal grains or grasses, which manifests itself under such multiplied forms as maize. From the little shrubby stalk that grows on the shores of Lake Superior, to the palmetto-like corn of the Mexican valleys; and from the tiny ears and flattened, closely clinging grains of the former, the brilliant rounded little pearl, or the thickly-wedged rice corn, to the magnificently elongated, swelling ear of the Kentucky, with its deeply-indented gourd-seed, it is developed in every grade of sub-variety. The kernels are long, round or flat, or shriveled like the sweet; and their color is white, yellow, blue, red or striated; yet each contains the same principles of nutriment, combined in somewhat different proportions, and contributes for equal weights, nearly in the same ratio, to the support of man and the lower orders of the animal creation.

The analysis of corn as given by Dr. Dana, is in 100 parts, of flesh-forming principles, (gluten and albumen) 12.60; fat forming, (gum, sugar, starch, oil, woody fibre,) 77.09; salts, 1.31; water 9. The yellow contains more oil or fatty matter than the white, and therefore yields a stronger or richer food. This quality gives greater intensity to the peculiarities of flavor; and by those not accustomed to its use, it is not relished so well as the white. This is shown by the preference given to the latter in England and Ireland, where it has been recently introduced as a staple article of food. The large proportion of oil in this grain, increases its tendency to rancidity, when exposed to a hot and moist climate, unless previously prepared to resist this influence by kiln-drying.

Besides the kinds in general cultivation in this country, varieties have been occasionally introduced from abroad, of a character so different, as almost to entitle them to the distinction of independent species. Such are *the Chinese tree corn,* bearing its slender ears at the extremities of several expanded branches; *the Egyptian,* with its millet-like head;

the Oregon, with its separate husk or envelope for every distinct kernel.   But if we narrowly watch the vagaries of nature, we shall detect deviations from the ordinary standards of our domesticated varieties, which approximate so closely to the most fanciful of the exotics, that we are compelled to believe, that all those which have hitherto come within our notice, originated from one common head; and that the peculiarities of every description, are owing to the difference of soil, climate and culture, and the carefully-cherished eccentricities of nature, aided by a skillful science or well practiced art.   It is needless to particularize the many popular kinds of corn under successful cultivation in this country.   They are found to vary with almost every degree of latitude and longitude; and there are not unfrequently, numerous kinds held in deservedly high estimation within a single district.   From these, there will be no difficulty in selecting such as will best repay the farmer's attention.

The Soil for Corn must be dry, rich and well-pulverized. Neither strong clay or poor wet lands will yield good crops of corn.   Land can scarcely be too rich for it; and the fresher and less fermented the manure applied to it, unless on light, sandy soils, the better it will be for the crop.   A great error is committed in raising corn, as with most of our tillage crops, from not having the soil sufficiently enriched; though this error is diminished in the case of such as will not bear an excess of manure.   Corn is a gross feeder, and necessarily ranges over a great space in search of food.   It has a large amount of stalk, leaves and grain to provide for in a few weeks, and its increase will be commensurate with the supply of food.

A clover ley or rich grass sod is an excellent preparation for corn, with the addition of manure when required.   But the manure should always be scattered broadcast, plowed and well harrowed in.   The roots will be certain to find it, and in consequence of its general diffusion and consequent gradual absorption by the crop, the development of the ear and grain will correspond with that of the stalk and leaves. When manured in the hill on poor soil, it comes forward rapidly, and this induces an extension of the roots and foliage, entirely disproportioned to the elements contained in the soil; and finding a support wholly inadequate to a corresponding maturity, the crop is limited to the overgrown stalks and leaves and a small proportion of grain.

The Selection of Seed should be made with the utmost

care, not only from the best varieties, but the best seed of the particular kind desired. Some of the choicest have been brought to their present perfection, by selecting only the earliest and largest ears from the most prolific stalks. This ought always to be done before the corn is gathered in the field, where there is an opportunity for comparison.

Hybridizing Corn, like that of other grain, is easily accomplished, and its results are marked and frequently beneficial. The probable identity in origin of all the varieties, is evinced by the rapid change exhibited in the most diverse kinds when their locality is changed. The small, early corn of the North, becomes the tall, later-maturing corn of the South, after being cultivated for two or three seasons in Louisiana; and the Oregon, with every kernel safely encased in its separate calyx, in the climate of New York, soon exchanges this partial covering for the more comprehensive husk. Similar changes are characteristic of every variety ever coming within our notice.

Preparation of Seed.—Repeated experiments have de monstrated the great utility of steeping corn for one or two days before planting, in a solution of saltpetre. This accelerates the growth of the plant, and is a protection against birds, squirrels and mice, and for a while it will keep off worms. An effectual remedy against these depredations, is to add half a pint of boiling tar to a peck of seed, stirring the corn briskly for several minutes. as the tar is added, till every kernel is thinly coated with it. This supersedes the necessity of the worse than absurd remedy of scare-crows. The crows and other birds are of great advantage to the farmer on all his fields, as they pick up numberless insects, grubs and worms, which infest the ground and destroy, or seriously injure the vegetation. Instead of driving them from the corn grounds, they should be enticed there, by every proper means; and by rendering the grain distasteful, their appetites are sharpened for the worms and insects, the less conspicuous, but more fatal enemies of the grain.

Planting.—Corn may be planted in hills three to five feet asunder, leaving from three to five stalks well spread in each hill, according to the kind of seed and quality of land Some plant in drills, but this is objectionable when raised for the grain, as the trouble of cultivation is greater, without increasing the yield. Thick planting gives fewer ears upon a stalk and those of less size. The time of planting at the North. is usually the first three weeks of May, depending much on the season. Late frosts will sometimes cut down the first leaves, without destroying the germ; but it is always best to defer planting till all apprehensions of it are removed. In the more southern States, earlier planting is desirable, and it is there put into the ground from February to April. To give regularity to the rows and facilitate after culture, the furrows for the seed should be struck out each way with the utmost exactness, and twice the corn planted that is necessary to remain. It requires to be covered about two inches deep. The surplus plants can be pulled up a the second hoeing, when all fear of injury is past. If the land is light, it should be laid flat before planting, and after this, rolled compactly.

Planting machines have been recently invented for putting in this grain, which greatly diminish the labor, while they perform the operation more perfectly. A light horse, or mule and boy can furrow and drop the seed, cover and roll, from eight to twelve acres per day; and with entire uniformity as to distance, depth of covering, and quantity of seed in each hill.

Cultivation.—The ground may be stirred when the plants first show themselves. This is most economically done with the cultivator or light plow, and if the operation be frequent and thorough, there will be little use for the hoe. Hilling or heaping the earth around the plants should always be avoided, except with very heavy soils, or such as are liable to an excess of moisture; in all other cases it should remain flat. Stirring the ground in dry weather, is peculiary beneficial to corn and all hoed crops. Some omit it then from fear of the escape of moisture, but its effect is precisely the reverse, for nothing so certainly produces friableness, porosity and unevenness in the soil; and this we have shown, under the heads of *soils* and *draining*, facilitates the admission and escape of heat, which inevitably secures the deposit of large quantities of moisture, even in the driest and most sultry weather. Corn and other crops that were withering from excessive drought, have been at once rescued from its effects, by a thorough use of the plow and cultivator. Well-drained, dark-colored, and rich porous soils will be found to suffer much less in drought, than others which lack these characteristics.

Harvesting.—If there be no danger of early frost, the corn may be suffered to stand till fully ripe; though if the stalks are designed for fodder, they are better to be cut when the grain is well glazed, and this should be done in all cases where frost is expected. Scarcely any injury occurs either to the leaf or grain, if the corn be cut and stooked, when both would be seriously damaged from the same exposure if standing.

The stalks of corn ought never to be cut above the ear, but always near the ground, and for this obvious reason. The sap which nourishes the grain, is drawn from the earth, and passing through the stem, enters the leaf, where a change is effected, analogous to what takes place in the blood when brought to the surface of the lungs, in the animal system; but with this peculiar difference, however, that while the blood gives out carbon and absorbs oxygen, plants, under the influence of light and heat, give out oxygen and absorb carbon. This change prepares the sap for condensation and conversion into the grain. But the leaves which thus digest the food for the grain are above it, and it is while passing downward, that the change of the sap into grain principally takes place. If the stalk be cut above the ear, nourishment is at an end. It may then become firm and dry, but it will not increase in quantity; while if cut near the root, it not only appropriates the sap already in the plant, but it also absorbs additional matter from the atmosphere which contributes to its weight and perfection.

Corn must be perfectly dried in the field, and after this husked and carried into an airy loft, or stored in latticed or open barracks. The stalks may be housed, or carefully stacked for fodder. Many of our western farmers allow both grain and stalks to stand in the field till wanted for use, when they are fed in an adjoining enclosure. This is a wasteful practice, and can only be justified by the very low price of grain. Where labor is not relatively too high, it is better to grind or crush the corn and cob, and cut the stalk; then mix all together, dampening and slightly salting the mixture some time before feeding it. Could a comprehensive machine be invented for grinding the whole mass of stalk, husk, cob and grain together, it would save much of the food, and the labor both in preparing and digesting it. When fodder is high, the stalks and leaves will repay the expense of cultivation.

Preparation of Corn for a distant market requires that the grain be not only well cured, sound and dry, but that it be properly *kiln-dried.* This expels the moisture, and destroys that vitality, which impels it to absorb dampness whenever exposed to it, as a preliminary aid to germination; thus carrying out that great law of reproduction, impressed by Deity on every organic structure, whether animal or vegetable, " whose seed is in itself." By the operation of kiln-drying, it becomes *mere matter divested of vitality*, and may then be carried into all climates with impunity.

Corn for Soiling.—Corn has recently been much cultivated for fodder, and for this purpose, the soil must be in high condition and well pulverized. It may be sown broadcast and harrowed in, at the rate of three or four bushels per acre. But a much better method is, to sow thickly in drills, and stir the ground with a light plow or cultivator. The sowing may be done early or late, though the first is most successful. It should be cut before the frosts touch it, and dried previous to housing. Several tons of excellent forage have been raised in this way, from a single acre. In a report to the Pedee Agricultural Society of South Carolina, it is asserted, that 138,816 lbs. of green corn stalks have been cut from one, acre in a season, weighing when dry, 27,297 lbs.

The Uses of Corn in this country are numerous. It is largely fed to fattening and working animals, but must be cautiously given to the latter, and especially in hot weather. It is extensively manufactured into high wines and whiskey, the consumption of which as a beverage, evinces a sad perversion of one of the best gifts of nature. It is converted into oil, molasses and sugar to a very limited extent; and is variously and largely applied to domestic uses. While green it is boiled or roasted in the ear; or it is cut from the cob and cooked with the garden or kidney bean, which forms the Indian *succotash*. When ripe, it is hulled in a weak ley, then boiled and known as *hulled corn*, a most convenient and acceptable dish in the frontier settlements, remote from mills; or it is parched over a hot fire, affording a delicious lunch, and a convenient provision for hunters, as *popped corn*. *Hominy* or *samp* is a favorite dish, and consists of corn coarsely ground and boiled in water ; and *hasty pudding* differs from this, only in being made of fine meal. The meal may be compounded with milk and eggs into *jonny-cakes*, puddings, griddles and other delicacies, universally esteemed for the table ; and when scalded and mixed with the flour of wheat or rye, it imparts additional sweetness to bread, while it scarcely diminishes its proportionate nutritive properties

RICE (Oryza sativa, Fɪɢ. 38).

This grain probably contributes directly to the support of a larger number of the human family than any other plant. In China, and nearly the whole length of the southern part of Asia; throughout the innumerable and densely populated islands of the Pacific and Indian Oceans; in the southern part of Europe, and a large extent of Africa; and through no inconsiderable portion of the North and South American continent and its central islands, it is extensively grown, and forms the staple food of the inhabitants. Rice requires a moist soil, and is much more productive when subject to inundation. A hot sun is also necessary to mature it; and as a result of these two essential conditions, its culture is limited to regions much more circumscribed than are allotted to wheat, maize, or some of the usually cultivated plants. I subjoin, from an excellent article on *rice and its cultivation*, addressed to the writer by Dr. Cartwright, a practical planter of Mississippi.

Fɪɢ. 38.

Varieties of Rice.—" Of these there are many, but I am induced to believe that they are all essentially aquatic. All the varieties, yet discovered, flourish best under the inundation system of culture; yield more to the acre, give less trouble, and require less labor. But each variety grows well on light, moist uplands without irrigation, when cultivated with the hoe or plow. The product, however, is so much less than by the irrigation system, and the labor of tillage so much more, that the upland producer never can compete successfully with the lowlander. The former may curtail his expenses by growing rice for domestic uses, but he cannot profitably, produce it for sale. Besides the ten-fold labor which rice on upland requires in comparison with that cultivated by the irrigation system, it cannot be sown thick enough to make a large yield per acre. Space must be left for the plow or hoe to till the rice, which is not necessary in those localities where it can be overflowed at will, and the water drawn off as occasion may require.

Cultivation of Lowland Rice.—The method pursued on the rice lands of the lower Mississippi, is to sow the rice broadcast about as thick as wheat, and harrow it in with a light harrow having many teeth; the ground being first well plowed and prepared by ditches and e nbankments for inundation. It is generally sown in March, and immediately after sowing the water is let on, so as barely to overflow the ground. The water is withdrawn on the second, third, or fourth day, or as soon as the grain begins to swell. The rice very soon after comes up and grows finely. When it has attained about three inches in height, the water is again let on, the top leaves being left a little above the water. Complete immersion would kill the plant. A fortnight previous to harvest, the water is drawn off to give the stalks strength, and to dry the ground for the convenience of the reapers.

A different method is practiced in the northern part of Italy. The seed is sown in April, previously to which it is soaked a day or two in water. After sowing, about two inches of water is let in upon the ground. The rice comes up through the water, which is then drawn off to give the plant strength, and after some days, is again let on. The rice is more apt to mildew under this practice, than our method, of letting the water on about the time the Italians draw it off.

The same measure of ground yields three times as much Rice as wheat. The only labor after sowing, is to see that the rice is properly irrigated; except in some localities where aquatic plants prove troublesome, the water effectually destroying all others. The rice grounds of the lower Mississippi produce about seventy-five dollars worth of rice per acre. The variety called the Creole white rice, is considered to be the best.

Cultivation of Upland Rice.—In the eastern part of the State of Mississipi, called the *Piney woods,* rice is very generally cultivated on the uplands. Although it cannot be made a profitable article of export, yet it affords the people of the interior an abundant supply of a healthy food for themselves, and a good provender for their cattle, and makes them independent of the foreign market. Unlike other kinds of grain, it can be kept for many years in a warm climate, without spoiling, by winnowing it semi-annually, which prevents the weevil and a small black insect that sometimes attacks it.

It is cultivated entirely with the plow and harrow, and grows well on the pine barrens. A kind of shovel plow drawn by one horse, is driven tl rough the unbroken pine-forest; not a tree being cut or belted, and no grubbing being necessary, as there is little or no undergrowth. The plow makes a shallow furrow about an inch or two deep, the furrows about three feet apart. The rice is dropped into them and covered with a harrow. The middles, or spaces between the furrows, are not broken up until the rice attains several inches in height. One or two plowings suffice in the Piney woods for its cultivation—weeds and grass, owing to the nature of the soil, not being troublesome. A similar method of cultivation obtains on the prairie land of the northwestern States. Rice, like hemp, does not impoverish the soil.* On the contrary, it is a good preparatory crop for some others, as Indian corn. The pine barrens of Mississipi would produce rice *ad infinitum*, if it were not that the land, after a few years, owing to the sandy nature of the soil, becomes too dry for it.

It has been ascertained by Arnal, that twelve pounds of wheat flour and two pounds of rice will make twenty-four pounds of an excellent bread, very white and good; whereas, without the addition of rice, 14 pounds of flour will only make 18 pounds of bread. Like other kinds of grain, rice adapts itself to the soil and climate, and particular mode of cultivation; but if the seed be not changed, or selected from the best specimens of the plant, it will ultimately degenerate. Thus in Piedmont, after a long series of years, the rice became so much affected with a kind of blight called the *brusone*, as to compel the Piedmontese to import fresh seed in 1829, from South Carolina. The American rice introduced into Piedmont, escaped the *brusone*, but it was several years before it adapted itself to the soil and climate.

Some years ago, a traveller, finding rice growing in great perfection on the mountains and highlands of Asia, particu-

---

* If this remark be limited to the *lowland* rice, we fully agree with it; as the water and the materials it holds, either in suspension or solution, and to which it is exposed through so long a period of its growth, afford the greater part of the nutritive matter appropriated by the plant. But if applied to *upland* rice, we must dissent *in toto;* for the rich, life-sustaining principles of this grain, draw largely on the soils where water is not present; for like the white grains, the wheat, oats, and barley, its narrow, grass-like leaves do not draw much from the atmosphere. The intelligent writer indirectly concedes this in the following sentence but one.   *The soil becomes too dry for it,* simply because it is exhausted of those vital, fertilizing principles, the salts and carbonaceous matters, which help to sustain the requisite moisture in the soil, and which is one of the beneficial results of their presence in it

larly Cochin China, named it *riz sec* or dry rice, and sent the seed to Europe, where many experiments were made with it. It yielded no better than any other kind of rice, and was found like all others to succeed best when inundated. The reason why it yielded so much more in Asia than in Europe can be readily accounted for, by the natural inundations it receives from the excessive rains during the monsoons.

No variety has been discovered which yields as much out of the water as it does in it. There are many localities in the United States, where the culture of rice by the irrigating system, would rather serve to make the surrounding neighborhoods healthy instead of sickly. It is generally admitted, that a given surface of ground completely inundated, is much less unhealthy than the same surface partially inundated, or *in transitu* between the wet and the dry state. Hence mill-ponds which partially dry up in the summer, are fruitful sources of disease. Some of the best rice is said to grow on the bottom of mill-ponds. Nothing more is necessary, than to make the bottom perfectly level, and then to overflow the whole surface just deep enough to keep the top leaves above water. As if to show that unhealthiness is not necessarily connected with the culture of this valuable grain, nature has imposed a law upon it, ordering that it should flourish better when overflowed with pure running water than with the stagnant waters of impure lakes and marshes.

There are two kinds of rice, which are said to succeed best on uplands, the long and the round. The former has a red chaff, and is very difficult to beat. The latter shakes out, if not cut as soon as ripe. They nevertheless succeed best under the inundation system of culture. In the eastern hemisphere, rice is cultivated as far North as the 46th degree of latitude. The climate of the United States is better suited to it than that of Europe, because our summers are hotter. In the northern part of China, the variety called the imperial rice, or *riz sec de la Chine* (the *oriza sativa mutica*), is more precocious than any other, is said to yield a heavy harvest, and to constitute the principal food for the people of that populous region. But it has succeeded no better in Europe than any other kind of rice.

The best rice lands of South Carolina are valued at five hundred dollars per acre, while the best cotton lands sell for a tenth part of that sum, proving that rice is more profitable than cotton. The profits of a crop should not so much be estimated by the yield per acre, as the number of acres a labor-er can till. After the land is properly prepared for inunda-tion, by levelling, ditching, and embankments, a single indi-vidual can grow almost an indefinite quantity of rice. Rice is no doubt ultimately destined to supersede cotton in a large portion of Mississippi and Louisiana."

The varieties of Rice most grown in South Carolina and Georgia, which have hitherto been the greatest rice-produc-ing States of the Union, are the *Gold-seed* rice, the *Guinea,* the *Common White,* and the *White-bearded.* There are several other varieties, but generally inferior to the foregoing. The best are produced by careful cultivation on soils suited to this grain, and by a careful selection of seed.

In 1839, South Carolina produced over 66,000,000 lbs.; Georgia, 13,400,000 ; Louisiana, 3,765,000 ; and North Caro-lina, 3,324,000, no other State producing one million pounds. Rice will keep for years uninjured, if allowed to remain in the chaff or husk as it is gathered, in which condition it is called *paddy.*

From the immense extent of our lowlands throughout the delta of the Mississippi, which, if subjected to the wet tillage of rice, may be considered of inexhaustible fertility, we may expect at some future day, to surpass every other portion of the globe in the quantity, as we now do in the quality of our rice.

MILLET (Panicum milliaceum).

This is the species of millet usually grown in the United States. In its form and the manner of bearing its seeds, the millet strongly resembles a miniature broom corn. It grows to the height of two to four feet, with a profusion of stalks, heads and leaves, which furnish excellent forage for cattle. From 60 to 80 bushels of seed per acre have been raised, and with straw equivalent to one or two tons of hay ; but an average crop may be estimated at about one third of this quantity. Owing to the great waste during the ripening of the seed, from the shelling of the earliest of it before the last is matured, and the frequent depredations of birds which are very fond of it, millet is more profitably cut when the first seeds have begun to ripen, and then harvest-ed for fodder. It is cured like hay, and on the best lands yields from two to four tons per acre. All cattle relish it, and experience has shown it to be fully equal to good hay.

Cultivation.—Millet requires a dry, rich, and well pul-verized soil. It will grow on thin soil, but best repays on

Fig. 38.

the most fertile. It should be sown broadcast or in drills from the 1st of May to 1st of July. If for hay and sown broadcast, 40 quarts per acre will be required; if sown in drills for the grain, eight quarts of seed will suffice. It will ripen in 60 to 75 days with favorable weather. When designed for fodder, the nearer it can approach to ripening, without waste in harvesting, the more valuable will be the crop.

INDIAN OR GRAND MILLET (*Sorghum vulgare*, Fig. 38).—This millet is much cultivated in Asia Minor, Egypt, Arabia, the West Indies, and elsewhere. It grows from four to six feet high, affording a large quantity of forage, and much seed or grain, which is known as Guinea corn. This is ground into flour and used by the laborers where grown. It is also an economical food for cattle, swine, and fowls. It is not raised to any extent in the United States, but might be advantageously introduced into the southern States.

BUCK-WHEAT, OR BEECH-WHEAT (*Polygonum fagopyrum*, Fig. 39), is a grain much cultivated in this country. It grows freely on light soils, but yields a remunerating crop only on those which are fertile. Fresh manure is particularly injurious to this grain. Sandy loams are its favorite soils, especially such as have lain long in pasture, and these should be well plowed and harrowed. It may be sown from the 1st of May to the 10th of August, but in the northern States, this ought to be done as early as June or July, or it may be injured by early frosts, which are fatal to it. It is sown broadcast, at the rate of three to six pecks per acre, and harvested when the earliest seed is fully ripe. The plant often continues flowering after this, and when the

Fig. 39

early seed is blighted, as is often the case, the plant may be left till these last have matured. As it is liable to heat, it should be placed in small stacks of two or three tons each, but it is better to thresh out the grain at once. If not perfectly dry, the straw may be stacked with layers of other straw, and when well cured, it will be a valuable fodder for cattle. Sheep will feed and thrive as well on this straw as on good hay.

Uses.—This grain is ground and bolted and the flour is much used for human food. Before grinding, the hull or outer covering should be removed. When thus prepared, the flour is as white and delicate in appearance as the best rye, it is equally light and digestible, and is scarcely inferior to wheat in its nutritive properties. The grain is used for fattening swine, but is most profitable when mixed with corn. Poultry thrive upon it. Buckwheat was formerly employed as a fertilizer, but for this object it is inferior to

the clovers, in all cases where the soil is capable of sustaining them. Its rapid growth will insure the maturing and turning under of two crops in one season. There are other varieties than the one specified, but none of equal value for general cultivation in this country.

CANARY GRASS (*Phalaris canariensis*, Fig. 40).—This, like the millet, is an annual, and is used like many other species of the family of grasses, both for the seed and forage. Its chief use, however, is as a food for the canary, and other feathered pets.

It is sown quite early in the season, in drills, 12 to 18 inches apart, at the rate of two or three pecks per acre, in a rich, well-pulverized loam ; lightly covered, and kept clear of weeds by the cultivator and hoe. It is cut when fully ripe, and allowed to remain for some time exposed to the dews or rain, to loosen the chaff, which otherwise is very difficult of removal.

FIG. 40.

CHAPTER VIII.

LEGUMINOUS PLANTS.

THE PEA (Pisum sativum).

THE pea, bean, tare or vetch, lupine, the clovers, and some other plants, are all embraced in the botanical order *Leguminosæ.* The pea is valuable for cultivation, not only for the table, but for many of the domestic animals. It is largely fed to swine, sheep and poultry. For the former, it should be soaked, boiled or ground. If land is adapted to it, few crops can be more profitably raised for their use. It ripens early, and when beginning to harden, they may be fed with the vines, and the animals will masticate the whole, and fatten rapidly.

The Soil.—The heaviest clays will bear good peas, but a calcareous or wheat soil is better. Strong lands produce the best crops, but these should be made so by manures previously applied, as the addition of such as are fresh, increases the growth of haulm or straw, and sometimes diminishes both the quantity and quality of the pea. When sown on a poor sward, the manure should be spread before plowing. A dressing of well-rotted manure increases the crop, and is a good preparation when intended to be followed by wheat.

Varieties.—Of these there are many. The earlier kinds are generally indifferent bearers, and their cultivation is limited almost exclusively to the garden. Of those for field culture, the marrow-fat are among the richest of the peas, and they are preferred for good lands. The small yellow are perhaps the best for poorer soils. There is a very prolific *bush-pea* grown in Georgia, bearing pods six or seven inches long, which hang in clusters, on a short upright stem. The pods are filled with a white pea, which is highly esteemed for the table, either green or dry. In that latitude, they bear two or three crops in one season.

Cultivation.—Peas should have a clean fallow or fresh, rich sod, well harrowed. They are not affected by frosts, and may be sown as soon as the ground is dry. This will enable them to ripen in season to plow for wheat. The are very liable to attack from the pea-bug, which deposits its egg in the pea while in its green state, where it hatches; and the worm, by feeding on the pea, diminishes its weight nearly one half. Here it remains through the winter, and comes out as a bug the following season. To avoid this pest, some sow only such seed as has been kept over two years, while others sow as late as the 15th to the 25th of May, which delays the pea till after the period of its attacks, but this latter practice seldom gives a large crop. It may be killed by pouring boiling water upon the seed, stirring for a few minutes, and then draining it off. Peas are sometimes sown in drills, but most usually broadcast, at the rate of two or three bushels per acre. It is better to plow them in, to the depth of three inches, and afterwards roll the ground smooth, to facilitate gathering. When sown in drills, they may be worked by the cultivator, soon after coming up. The growth is promoted by steeping the seed for twenty or thirty hours in urine, and then rolling in ashes or plaster.

Harvesting is accomplished by cutting with the sickle or scythe, or what is more expeditious, when fully ripe so that the roots pull easily, with the horse rake. When thus gathered into heaps and dried, they may be threshed, and the haulm carefully stacked for sheep fodder. If this is secured in good condition, cattle and sheep will thrive upon it. Peas are frequently sown with oats, and when thus grown, they may be fed to sheep or horses as harvested or threshed, or made into meal for swine.

The Cow or Indian Pea, frequently called *the Stock Pea.* *The Southern Bean* would be a more appropriate name for it, as it is grown exclusively in the southern States. It is a desirable crop, either as a fertilizer, or as food for domestic animals. Its long vines and succulent leaves, which draw much of their substance from the air, and its rapid and luxuriant growth, particularly adapt it to the first object; while its numerous and well-filled pods, and its great redundancy of stem and leaf, afford large quantities of forage. This is improved for cattle, when harvested before the seed is fully ripe. It is sown broadcast, in drills, or hoed in among corn, when the latter is laid by for the season. If in drills, it may be cultivated in its early stages by the plow, shovel-harrow, or cultivator. It can be cut with the scythe, or drawn together with a heavy iron-toothed harrow, or horse rake, as with the common pea. It requires a dry, mellow soil, and is well suited to clays

THE BEAN (Phaseolus vulgaris).

The bean is often a field crop in this country, and espe cially in the northern and middle States. It is principally used for the table, either green or dry. It is a palatable and highly condensed food, containing much in a small compass. In proportion to its weight, it gives more nutriment than any of the ordinary vegetables; according to Einhof, yielding 84 per cent. of nutritive matter, while wheat gives only 74. It has, in common with the pea and vetch, though in a greater proportion, a peculiar principle, termed *legumin,* which is analogous to *casein,* the animal principle in milk. This is convertible into cheese, and in its nutritive properties, it is essentially the same as the *fibrin* of lean meat, the *albumen* of eggs, and other animal matters. There is no vegetable we produce, which so nearly supplies the place of animal food, as the bean.

Soil.—The bean is partial to a quick, dry soil. Too great strength of soil, or fresh manuring, gives a large quantity of vine, without a corresponding quantity of fruit.

Cultivation.—The land should be finely pulverized, and if at all inclined to wet, it should be ridged. Beans are tender plants and will not bear the slightest frost, and as they grow rapidly, they will be sure to ripen, if planted when this is no longer to be apprehended. The seed is exposed to rot if put into the ground in cold, wet weather, and the land should, therefore, be previously well warmed by the sun. The bush beans are the only kind used for field planting, and of these there are several sub-varieties. The long garden beans, white, red or mottled, are great bearers, of fine quality, and early in maturing. This is important, when other crops are to succeed the same season. They are usually planted in hills, about two feet apart, and also in drills, and covered with two inches of fine earth. They have been sown broadcast, on clean, dry soils, and produced largely. When planted in drills, from five to eight plants should be left in each, according to their proximity; or if in drills, they need from six to eight pecks of seed to the acre.

Harvesting.—When the beans are fully formed, and there is danger of frost, pull and throw them into heaps, in which condition the frost scarcely affects them. If the ground is not wanted for other uses, they may stand till the latest pods assume a yellow color. They are pulled with ease when the plant is mature, as the fibres of the root are by that time dead. This is more quickly accomplished with an iron hook-rake, or if the stalks are partially green, they can be mown. If the vines are not dry, let them remain for a while in small heaps, and afterwards collect in larger piles, around stakes set at convenient distances, with the roots in the centre and secured at the top by a wisp of straw. When well dried, thresh, clean and spread them till they are quite free from dampness.

Uses for Farm stock.—The straw or haulm is an excellent fodder for sheep, and it ought always to be stacked for their use. Sheep are the only animals which eat them raw ; and for them, no species of grain is better suited than the bean, when fed in moderate quantities. Swine, cattle and poultry, will thrive on them when boiled. Sixty bushels have been raised on an acre, worth from one to two dollars per bushel.

THE ENGLISH FIELD BEAN (Vicia faba).

Is cultivated under many varieties in Europe, and particularly in Great Britain, as a field crop for the use of horses and other animals. Among these are *the Windsor, the tick, the long pods and others.* Arthur Young prefers "the common little horse-bean as being more generally marketable." I have raised several of these varieties, and although entirely successful, have found them less adapted to our climate and agriculture, than the ordinary crops. They prefer a strong clay, or loamy clay soils.

THE TARE, VETCH OR FITCH (V. sativa, Fig. 41).

Of this there are two kinds, the winter and the spring, both of which are hardy and productive. It is deemed an important crop in Europe, where it is much cultivated for green fodder or soiling, and frequently it is used as pasturage, or cut and cured for hay. It is partial to clay, but grows indifferently on any rich soil which is not too dry. It is sown broadcast or in drills, but generally the former, on well-pulverized lands, and covered with the harrow, demanding no after attention except the extermination of weeds. Tares have

Fig. 41.

hitherto been little grown in this country, but in certain soils and situations, they may be introduced as a substitute for clover, where, from any cause, the latter does not grow successfully. All domestic stock are fond of them.

THE PINDAR, GROUND PEA, OR PEA-NUT (Arachis hypogœa).

This is a legumen and is cultivated with profit in the southern States, on light, loamy or sandy lands, where it yields from 30 to 60, or even 80 bushels per acre, besides furnishing much haulm for forage. It is planted in hills, or sown in drills four to five feet apart, and worked with a light plow or cultivator, immediately after the plants show themselves above ground. They soon overspread the whole surface. When properly matured, the roots are loosened by a fork and pulled up by hand, and after curing, are put under cover for winter's use. They contain a large quantity of oil of a superior quality and flavor, which is suitable for the table and various purposes in the arts.

The peat-nut is in high repute for its fattening qualities, when fed to stock. Swine are particularly partial to them; and if allowed to run on a field containing both them and corn, they will remain among the pea-nuts till entirely exhausted, resorting to the grain occasionally, for a change of food. They can lie in the ground all winter, uninjured by frosts or rains. They are much used for human food after drying and baking.

THE WHITE LUPINE (Lupinus albus, Fig. 42).

THIS plant is sometimes raised in southern Europe, where the seed is used as human food. It was cultivated by the Romans, and others among the ancients, for the same purpose. It is frequently used as a forage plant, for which purpose the whole plant is cut and fed green, or cured as hay. It is sometimes made use of as a fertilizer, for which it is well adapted It requires a similar soil and cultivation with the pea

Fig. 42.

CHAPTER IX.

ROOTS.

THE POTATO, (Solanum tuberosum).

THE potato is a native of the American Continent. It is found in a wild state both in Buenos Ayres and Chili, and was probably discovered in the same condition by the early settlers of North America. It was supposed to have been taken into Spain and Italy, early in the 16th century, by Spanish adventurers, as it was cultivated in those countries in 1550. In 1588, it was introduced into Vienna from Italy; and also into England, probably as early as 1586, by the colonists of Virginia, who were sent out by Sir Walter Ra-'eigh. On its first introduction into Europe, it was considered a delicacy; and it is not until within a comparatively recent period, that it has found its way into both continents as an article of agricultural attention, and an almost indispensable food for man and beast.

Varieties.—These are almost illimitable. In form they are round, oblong, flat and curved or kidney-shaped; they vary in size from the delicate lady-finger to the gigantic blue-nose; their exterior is rough or polished, and of nearly every hue, white, yellow, red, and almost black; and the surface is sometimes smooth and even, with the eye scarcely discernible; or deeply indented with innumerable sunken eyes, like the Rohan and Merino. The interior is equally diversified in color; and is mealy, glutinous or watery; sometimes pleasant and sometimes disagreeable to the taste. They likewise differ in ripening earlier or later, and in being adapted in some of their varieties to almost every peculiarity of soil.

New kinds are produced at pleasure, by planting the seed found in the balls. The tubers obtained in this way, are small the first season, but with careful culture, will be large enough the second year to determine their quality, when the best may be selected for propagation. The earliest are easily designated by the premature decay of the tops. The varieties may also be increased from the seed by hybridizing, or impregnating the pistils of one flower by the pollen from another; and in this way, some of the best and most valuable kinds have been procured. Such as have no flowers are more productive of tubers, as there is no expenditure of vitality in forming the seed. They may be compelled to flower, by removing the small tubers from the roots as they form.

The best Soil for Potatoes is a rich loam, neither too wet nor too dry. Cool and moist soils, like those of Maine, Nova Scotia and Ireland, and especially, if in rich, fresh sod, give the best flavored potatoes, and such as are the least liable to disease. A calcareous soil yields good potatoes, and generally sure crops, and where there is little lime in the soil, it should be added. Ashes, salt and gypsum are excellent manures, and in certain instances, have astonishingly increased the product. Crushed bones also greatly improve a potato soil. Fresh manures will often affect the taste of the potato unpleasantly, and when necessary to apply them, they should be scattered broadcast and plowed in.

Select such seed as experience has decided is best adapted to the soil, and the use for which they are to be appropriated. Some are careful to cultivate the most mealy for the table, and plant those which give the greatest yield for their cattle. This is sometimes mistaken policy, as what are best for man, are generally best for cattle; and although the farmer may get a much greater weight and bulk, on a given quantity of land of one kind, these may yield a less quantity of fat and flesh-forming materials, than those afforded by a smaller quantity of some other variety. Experiment has shown, that of " three varieties grown in Scotland, in 1842, the *cups* gave $13\frac{3}{4}$ tons per acre, containing $2\frac{9}{10}$ tons of starch; the *red dons* yielded $14\frac{1}{4}$ tons and $1\frac{9}{10}$ of starch; the *white dons*, $18\frac{1}{2}$ tons, and $2\frac{4}{10}$ of starch, and the kidney has even given as much as 32 per cent. of starch."—(*Johnston.*)

There is also a difference in the relative proportions of gluten. The potato contains in its new and ripe state, about $2\frac{1}{4}$ per cent., which diminishes by long keeping. It is important for this, as for an indefinite number of other practical matters, to have agricultural laboratories of unquestionable reliability, where the errors of superficial observation may be detected; and where the real superiority of one product over another, and their variations induced by soils, manures and treatment, may be established beyond the possibility of a doubt.

Planting.—To produce abundantly, potatoes require a fertile soil, and if not already sufficiently rich, spread manure on the surface before plowing. If a tough sod, plow the preceding fall ; or if friable, it may be done just before planting ; but in all cases, the land must be put in such condition as to be perfectly loose and mellow. Hills are the most convenient for tillage, as they admit of more thorough stirring of the ground with the cultivator or plow. Medium size, uncut potatoes have been ascertained, from numerous experiments, to be the best for planting, but when seed is scarce, it is sometimes economical to divide them. Two potatoes should be placed in each hill ; or if in drills, they should be planted singly, ten inches apart. The distance both of hills and drills must depend on the strength of the soil and the size of the tops, some varieties growing much larger than others. Cover with light mold to the depth of four or six inches, and if the soil be light, leave the ground perfectly level ; if cold, heavy or moist, let the hill be raised when finished. Subsoil plowing is of great benefit to potatoes, as to most other crops, whenever the soil will justify its use. The sets cut from the seed-end, give a much earlier crop than those from the root.

Cultivation.—When the shoots first appear above the ground, run the plow through them and throw the earth well to the plants ; and no injury results, if the tops are partially, or even entirely covered. The hoe is scarcely required, except to destroy such weeds as may have escaped the plow. The ground should be several times stirred before the tops interfere with the operation, but never after they come into blossom. Very large crops have been produced by top-dressing with compost, or well-rotted chip manure, soon after the plants make their appearance. This is carried to the field and spread from a light, one horse cart, the wheels passing between the rows ; but the same results would probably be attained, by placing the land in the best condition before planting, if followed by the nicest cultivation afterwards. There is some gain to the crop, when the buds are plucked before they come into blossom.

Harvesting and Storing should not be commenced until the tops are mostly dead, as the tuber has not arrived at full maturity before this time. They may then be thrown out of the hills by a double mold-board plow, or by a potato hook, or some other hand implement. They ought not to be exposed to the sun for any length of time, but may dry on

8

the surface in a cloudy day, or be gathered into small heaps, with some of the tops spread over them, until freed from the surface moisture, when they may be stored. Those selected for seed, should be placed in small piles in the field ; or in thin layers in a cool, dry place in the cellar, where the air is excluded and no heating or injury can occur. Those intended for winter consumption, may be put in dry bins or barrels in the store-room, and covered with straw and dry sand, or loose earth, to prevent the circulation of air. Such as are not wanted till the following spring, may be kept on the field, if there be not sufficient room in the cellars. It is better when thus stored, that they occupy an excavation on the north side of a hill, in a porous soil. If shaded by trees, it will tend to shield them longer from the heat of the sun, and they may thus be kept till June, before opening. They are generally stored in the level field, in an excavation one or two feet deep, four or five wide, and of any length required. They are piled as high as they can be conveniently ridged up, then covered with straw, carefully placed over them like shingles on a house, and covered lightly with earth till the severe frosts, when they should be adequately covered to protect them from rains and frost during winter. A partial heating and sweating take place soon after storing, and till this is complete, a loose covering of straw is all they require. A ditch lower than the base must encircle the heap when the soil consists of clay, from which an outlet conducts away all the water, as any left upon them will inevitably produce decay.

Diseases.—The potato has long been subject to the *curl*. From numerous experiments made in Scotland to avoid this disease, it has been found, that seed from potatoes which were gathered before fully ripe, gave a much better and surer crop. It would be well to try the experiment in this country, where there is any deficiency of product from want of full and healthy development. Potatoes are also affected by the *scab and grub*, against whose attacks there is no remedy, unless in a change of seed and locality.

The rot has for several years produced serious and increasing injury to the potato crop, threatening starvation in Ireland, and causing great loss and suffering in several other countries in Europe. Its effects have also been extensively felt in the United States. Numerous and scientific examinations have been made on the subject. The proximate cause is supposed to be a fungus, but what are the reasons for its ate rapid extensic n, and the remedy for its ravages, ɪave not yet been satisfactorily ascertained.

Preventives of Rot.—Under the following circumstances rot has not appeared, when adjoining fields have been nearly destroyed by it. 1. By using unripe seed, or seed which has been exposed to the sun, light and air, and well dried for ten days after digging, and afterwards stored in a dry place in small parcels, where air is excluded till the moment of planting. 2. By the use of lime, some of which is placed in the hill and the potatoes dusted with it, and also from the use of charcoal and salt, gypsum or other salts. 3. By the absence of fresh barn-yard manure, or if used, by adding largely of lime or saline manures. 4. The use of fresh sod, which has long been untilled. This has been found more efficacious than any other preventive, although it has occasionally failed. The sod may be plowed in the fall, or left till late in May or early in June, when it has a good coating of grass; then turn over the ground and furrow it lightly, to receive the seed without disturbing the sod. Or they may be planted by using a sharpened stake three inches in diameter, with a pin or shoulder ten inches from the bottom, on which the foot may be placed for sinking the holes. These should be made between the furrow slices at the proper distance for drills, and a single potato placed in each, which may be covered with the heel. 5. Sound, early varieties, early planted, have also escaped. I have thus secured a good yield, almost wholly free from disease ; and even those affected did not appear to communicate disease to others. It has also been found that some very late planted have escaped rot ; and if it be an epidemic, it may be, that both by early and late planting, the peculiar stage of vegetation when the fungus appears, is in a great measure avoided. But the investigations on this important subject are still in their infancy, and nothing has thus far been ascertained, which can be justly considered as having determined principles of universal application ; yet it is to be hoped that the zeal, intelligence and general interest which are now combined for this object, will ere long detect, what has hitherto evaded the severest scrutiny of scientific research.

Arresting the disease has in some instances been successful, by mowing off the tops when they are found defective. This practice would be injurious to healthy plants, but may be adopted, like that of cutting grain when struck by rust, if it will secure even a part of the crop. When disease appears in such as are dug, they should be carefully sorted and the sound ones well dried, then placed separately in layers and covered with ashes, burnt clay, or fine dry mold. These act as absorbents of moisture, and prevent contagion from such as may be imperceptibly affected. They may also be cut in slices and dried, or crushed and the farinaceous part extracted. By this means the potato will be made to yield nearly all its nutriment. It is found that this disease affects the tissues (the nitrogenized or albumenous part) of the potato only; and for this reason, potatoes which have not been too long or too deeply injured, will yield nearly their full amount of fat for animals or starch for the manufacturer.

Uses.—Besides being an almost indispensable vegetable for the table, potatoes are boiled and mixed with flour or bread, to which they impart a desirable moisture and an agreeable flavor. They are sliced, dried and ground, and much used in Europe as flour, and by the confectioners in their various products. They are also manufactured into tapioca, and when nicely prepared, it is scarcely distinguishable from that of the manioc. In all of these and some other forms, they enter into consumption as human food. They are also used in large quantities by the manufacturers of starch; to some extent for distilling; and in a less degree for making sugar. The refuse of the pulp, after extracting the starch and the liquor drained from it, are used for cleansing woolens and silks, which they effect without injury to the color. But by far the greatest use of potatoes in this country, is for stock-feeding. They are eaten with avidity by all the brute creation, either cooked or raw. For cattle and sheep, they are equally nutritious in either condition. For horses, they are improved by steaming or baking. Swine and most poultry will subsist on them raw, but will fatten on them only when cooked. Their good effects are much enhanced by mixing with meal when they are hot, which partially cooks it.

THE SWEET POTATO (Convolvulus batatas, Fig. 42)

Is a root of very general growth, in the southern, and it is much cultivated in the middle sections of the United States. It is scarcely surpassed by any esculent for the table. and it is greadily eaten, and with great advantage, by every species of stock.

Soil.—A dry, loamy soil, inclining to sand, is best for them; and this should be well manured with compost scattered broadcast, before working the ground, and thoroughly pulverized by repeated plowing and harrowing. It should then be thrown into beds four feet wide with the plow, and in the centre of these, strike a light furrow to receive the seed, if the soil is dry, or plant it on the surface, if moist. The use of a subsoil plow in the beds before being thrown up. is of great benefit to the plant.

Fig. 42.

Cultivation.—When the season is sufficiently long to mature them, the potato may be most conveniently planted, by cutting the seed into slips, and laying them six or eight inches apart in the place where they are to mature. Large potatoes divided into pieces of a proper size, are better for seed than small ones uncut. These should be covered about two inches with light mold. When they begin to sprout, the plow may be run close to the rows on either side, to remove the earth and allow the full benefit of the sun and air to the roots, and as the plant advances in its growth, the earth may be gradually restored to them by the plow and hoe. Where the vines are so large as to be injured by the plow, the hoe alone should be used. The hill or drills may then be made broadly around the plants, hollowing towards them, to afford a full bed of rich, mellow earth, and to retain the rain which falls. They are fit for gathering when the vines are dead.

When the season is short or early potatoes are wanted, plant on a hot bed, made of warm manure, with a covering of four inches of fine mold.   After splitting the potatoes, place them on this and cover with three inches of light earth.   As the sprouts appear, draw and transplant them after a rain, in the same manner as before suggested with the roots.   When early vegetated, a bushel of seed will, in this manner, supply plants for an acre.

The preservation of the Sweet Potato through the winter is often difficult.   A careful seclusion from air and light, and the absence of frost and absolute dryness seem to be essential to their preservation.   They are frequently kept, by piling in heaps on dry earth, which are still more secure with a layer beneath of corn stalks or dry pine boughs, six or eight inches deep.   On this, pack the roots in piles six feet in diameter.   Cover with corn stalks and dry earth, and protect this with a roof of boards, and a ditch deep enough to carry off all water.   There must be a hole at the top, slightly stopped with straw, to permit the escape of heated air, and to preserve uniformity of temperature.   There are numerous *varieties* of the sweet potato, white, red, yellow, &c.   They yield from 200 to 300 bushels per acre, and under favorable circumstances, sometimes double this quantity.

THE TURNEP (Brassica rapa).

The flat English Turnep was introduced into this country with our English ancestry, and has ever since been an object of cultivation.   When boiled, it is an agreeable vegetable for the table.   Its principal value, however, is as a food for cattle and sheep, by which it is eaten uncooked.   Its comparitive nutritive properties are small ; but the great bulk which can be raised on a given piece of ground, and the facility and economy of its cultivation, have always rendered it a favorite with such farmers, as have soil and stock adapted to its profitable production and use.

The proper soil for it is a fertile sand or well-drained loam.   Any soil adapted to Indian corn will produce good turneps.   But it is only on new land, or freshly-turned sod, that they are most successful.   An untilled, virgin earth, with the rich dressing of ashes left after the recent burning of accumulated vegetable matter, and free from weeds and insects, is the surest and most productive for a turnep crop.   Such land needs no manure.   For a sward ground, or clover ley, there should be a heavy dressing of fresh, unfermented manure, before plowing.

Cultivation.—Turneps are sown from the 15th of June to the 1st of August. The first give a greater yield; the last, generally a sounder root, and capable of longer preservation. The ground should be plowed and harrowed immediately before sowing, as the moisture of the freshly-turned earth insures rapid germination of the seed, which is of great importance to get the plants beyond the reach of insects as soon as possible. They may be sown broadcast, at the rate of one or two pounds per acre, and lightly harrowed, or brushed and rolled; or it is better that the seed be sown in drills, when a less quantity will suffice. A turnep drill will speedily accomplish the furrowing, sowing, covering and rolling at a single operation. The crop will be materially assisted by a top dressing of lime, ashes, and plaster, at the rate of fifteen or 20 bushels of the first two, and one and a half to three bushels of the last per acre. When the plants show themselves and the leaves are partially expanded, the cultivator or hoe may be freely used, stirring the ground well, and exterminating all weeds.

Ruta-Baga or Swedes Turnep.—The introduction of this is comparatively recent, and it proves to be more worthy of attention than the English or white turnep. It will grow in a heavier soil, yield as well, give a richer root, and it has the great advantage of keeping longer in good condition; thus prolonging the winter food of cattle when they most need it.

Cultivation.—It is usually planted after wheat or corn; but if a virgin soil or old pasture sward is chosen, it will materially lessen its liability to insects and other enemies. It is generally sown in drills, about two feet apart; and on heavy lands, these should be slightly ridged. The plants must be successively thinned, to prevent interfering with such as are intended to mature, but enough should remain to provide for casualties. Where there is a deficiency, they may be supplied by transplanting during showery weather. They should be left six or eight inches apart in the drills. The Swede turnep is a gross feeder, and requires either a rich soil or heavy manuring; though the use of fresh manures, has been supposed to facilitate the multiplication of enemies. Bones, ground and drilled in with the seed, or a dressing of lime, ashes, gypsum and salt, are the best applications that can be made. The Swede should be sown from the 20th May to the 15th June, and earlier than the English turnep. as it takes longer to mature; and two or three weeks more of growth, frequently adds largely to the product. **An** early sowing also gives time to plant for another crop in case of failure of the first.

Enemies.—The turnep is exposed to numerous depredators, of which the turnep flea-beetle is the most inveterate. It attacks the plant as soon as the first leaves expand, and often destroys two or three successive sowings. The black caterpillar, slugs, wire-worms, and numerous other insects, grubs and aphides prey upon and greatly diminish the crop.

Remedies have been tried to an almost indefinite extent, but none hitherto, with more than very partial success. Liberal sowing and rapid growth best insure the plant from injury; and to effect this, the seed should be plentifully sown in a rich soil, and if possible, when the ground is moist. Before sowing, the seed should be steeped in some preparation, which experience has shown will the most quickly develop the germ. Solutions of the nitrates or sulphates, urine, soot-water, liquid guano, or currier's oil, impregnate the first leaves with substances distasteful to their early enemies, and thus a short respite from their attacks will be secured. Gypsum, ashes, bone-dust and poudrette drilled in with the seed, are excellent forcers for the young roots. Charcoal dust applied in the same way, has been found to increase the early growth from four to ten-fold. When the fly or bug is discovered, the application of lime, ashes or soot, or all combined, should be made upon the leaves, while the dew or a slight moisture is on them. This leads the young plant along, and kills such enemies as it reaches. Stale urine, diluted sulphuric acid, (oil of vitrol,) and other liquid manures will have the same effect. Ducks, chickens, young turkeys and birds will devour innumerable quantities, and their presence should always be encouraged not only on this, but on most of the fields. Dragging the surface with fine, light brush, will lessen the slugs and insects. The ground should be plowed just before winter sets in, which exposes the worms and the larvæ of insects to the frost, when they are unable to work themselves into a place of safety. The seed should not be planted on ground recently occupied by any of the order of plants *cruciferæ*, (cabbage, radish, mustard, charlock and water-cress), as they all afford food for the enemies of turneps, and thereby tend to their multiplication.

Harvesting may be deferred till the approach of cold

weather; and in those sections of the country not affected by severe frosts, when on dry soils, they may be allowed to winter on the field. Otherwise, they should be pulled during the clear autumnal weather. This is accomplished most expeditiously with a root hook, which is made with two or more iron prongs attached to a hoe-handle. The use of a bill hook or sharp knife will enable the operator to lop off the leaves with a single blow, when they are thrown into convenient piles, and afterwards collected for storage.

The Storing may be in cellars or in heaps, similar to potatoes, but in a cooler temperature, as slight heat injures them, while frost does not. If stored in heaps, one or more holes should be left at the top, which may be partially stopped by a wisp of hay or straw, to allow the escape of the gases which are generated.

The feeding of Ruta-Bagas to cattle and sheep is always in their uncooked state. They are better steamed or boiled for swine; but food for these, should be sought from the more fattening products of the farm. They may be fed to horses in moderate quantities, but they cannot be relied upon for them, as they are too bulky for working animals. Their place is much better supplied for horses, by the carrot or potato. Their true value is as food for store and fattening cattle, milch cows and sheep, as they furnish a salutary change from dry hay; being nearly equivalent as fodder, to green summer food. They should be washed before feeding, if too much dirt adheres to them; but if grown on a light soil, the tap roots lopped off, and otherwise properly cleaned, they will not require it. They may be sliced with a heavy knife, or more summarily cut up while lying on the barn-floor, with a sharp spade, or root slicer, which is made with a socket handle and two blades crossing each other in the centre at right angles, or by some of the numerous improved cutting machines. With an abundance of turneps and a small supply of straw, hay may be entirely dispensed with for cattle and sheep, except during very cold weather. Many of the best English breeds, are kept exclusively on turneps with a little straw, till ready for the shambles.

The varieties of turneps are numerous. After selecting such as will give the largest crop of the most nutritous roots, the next object in the choice of particular varieties, should be to adapt them to the most economical use. Some will keep much longer than others, and if wanted to feed late in the season, it may be necessary to take a variety, intrinsically

8*

less valuable than another, which must be earlier consumed. The English turnep should be first fed, as it soonest
wilts and becomes pitt y; then follow with the others according to their order of maturity and decay. The leaves yield
good forage, and if unmixed with earth may be fed green or
dry to cattle.

The value of turneps to this country is trifling in comparison with that of many parts of Europe. In Great Britain
alone, this value probably exceeds one hundred millions of
dollars annually. But its culture here is much less desirable;
as our drier climate and early and severe winters are not as
well adapted to its production, and economical preservation
and feeding as those of England, and its numerous enemies
render it an uncertain crop. These objections are increased
by the important fact, that it enters into competition with
Indian corn, which generally gives a certain and highly remunerating return. It may sometimes, however, take the
place of corn with advantage; and the turnep or some of the
other roots should always occupy a conspicuous place as a
change, in part, for the winter food of cattle and sheep.

THE CARROT (Dracus carota).

This is one of our most valuable roots. It is a hardy, easily-cultivated plant; it grows in almost every soil, and is next
to the potato in its nutritive properties.

The soil which best suits it is a fertile sand or light loam;
but it will grow on such as are more tenacious, if well
drained, and deeply worked. The success of this and the
parsnep, depends much on the depth to which their roots
can reach. Deep spading or subsoil plowing is, therefore,
indispensable to secure large crops, and the ground should
be thoroughly pulverized. Barnyard manures, composted
with the different salts or ashes, or chip dung, are best for
them. It is desirable to have the manures well rotted, for
the double object of killing obnoxious seeds, and mixing intimately with the soil.

The varieties chiefly used for field culture are the long
red, the orange, and white Belgian. The last, under favorable circumstances, attains huge dimensions; and from its
roots growing high out of the ground, it is supposed to draw
more of its nourishment from the air, and consequently, to
exhaust the ground less, while it is more easily harvested.
But it is considerably below the others in comparative
value

Planting.—The carrot should be sown in drills, 16 to 20 inches apart, when the ground has become warm and dry. The seed is best prepared by mixing with fine mold or poudrette, and stirring it well together to break off the fine beards; then sprinkle with water and allow it to remain in a warm place, and occasionally turn it to produce equal development in the seed. It may remain 10 or 15 days before sowing, and till nearly ready to sprout. It then readily germinates, and does not allow the weeds to get the start. The frequent use of the cultivator and entire cleanliness from weeds, are all that is necessary to insure a crop; unless it be convenient to give it a top dressing of liquid manure, which the Flemings always do, and which no crop better repays. Two pounds of good seed will sow an acre. Any deficiency of plants may be supplied by transplanting in moist weather. Six inches is near enough for the smaller kind to stand, and eight for the larger. They are subject to few diseases or enemies, excepting such as can be avoided by judicious selection of soil and careful tillage.

The harvesting may be facilitated by running a plow on one side of the rows, when the roots are easily removed by hand. The tops are then cut and the surface moisture upon the roots dried, when they may be stored like turneps and potatoes. They ought to be kept at as low a temperature as possible, yet above the freezing point. On the approach of warm weather, they will sprout early if left in heaps; and if important to preserve them longer, the crown should be cut off and the roots spread in a cool, dry place.

Uses.—Carrots are chiefly grown for domestic stock. Horses thrive remarkably on them, and some judicious farmers feed them as a substitute for oats. But their intrinsic value in weight, for their fat and flesh-forming properties, is less, in the proportion of about five to one. For their medicinal properties, however, and the healthful effects resulting from their regular, but moderate use, they would be advantageously purchased at the same price as oats, or even corn, if they could be procured no cheaper. They are good for working cattle, and unsurpassed for milch cows, producing a great flow of milk, and a rich yellow cream. Sheep and swine greedily devour them, and soon fatten, if plentifully supplied. The Dutch sometimes grate them, and with sugar and salt, make a pickle for their choicest table butter. They are also employed in distilling. The average yield, on good land, may be estimated at about 300 bushels of the smaller, and 450 of the Belgian or white, per acre ; but with extra cultivation, 1,000 bushels of the last have been raised.

THE PARSNEP (Pastinaca sativa).

The parsnep is frequently cultivated as a field crop, and it is nearly equal to the carrot in its value. *The soil may be heavier for parsneps than for carrots*, and they will even thrive on a strong clay, if rich, well pulverized and dry. Large crops can only be obtained on deeply fertile and well pulverized soil. They should be sown early, as frosts do not affect them, and they require a long time to come to maturity. Drilling, at a distance of 20 inches apart, is the proper mode of planting, and they should be thinned to a space of six or eight inches. It requires four or five pounds of seed per acre, which must be of the previous year's growth, as older does not readily vegetate. No preparation of the seed is necessary. The subsequent cultivation is similar to that of carrots, and they will generally yield more under the same circumstances of soil and tillage. They are little subject to disease or enemies.

The best variety for field culture, is the Isle of Jersey. The gathering should be deferred till the following spring, unless wanted for winter's use ; as they keep best in the ground, where they are uninjured by the intensest frost. But particular care must be observed in allowing no standing water on them, or they will rot. When taken up in the fall, the roots should neither be trimmed nor broken, nor should the tops be cut too near the root. They must be stored in a cool place and covered carefully with earth, as exposure to air or even moderate heat wilts them.

Uses.—The parsnep is one of our most delicious table vegetables. It is an excellent food for swine, either raw or cooked, and for cattle, milch cows and sheep, it is highly prized. Qualey says, "it is not as valuable for horses, for though it produces fat and a fine appearance, it causes them to sweat profusely ; and if eaten when the shoot starts in the spring, it produces inflamation in the eyes and epiphora, or weeping." The leaves of both carrots and parsneps are good for cattle, either green or dried. Gerarde, who wrote in 1596, says, "an excellent bread was made from them in his time." They have also, like the carrot, been used for distillation, and are said to affo-d a *very good vinous beverage*

THE BEET (Beta).

There are but two varieties of the beet in general use for the field, *the Sugar beet and Mangold-wurzel,* both of which have several sub-varieties. They are of various colors, red, pink, yellow, white or mottled, but color does not seem to affect their quality. The conditions under which they grow are similar. Beets do well in any soil of sufficient depth and fertility, but they are perhaps, most partial to a strong loam. If well tilled, they will produce large crops on a tenacious clay. We have raised at the rate of 800 bushels per acre, on a stiff clay, which had been well supplied with unfermented manure. The soil, cannot be made too rich ; and for such as are adhesive, fresh or unfermented manures are much the best.

The planting should be in drills, 20 to 24 inches asunder, at the rate of four to six pounds of seed per acre, buried not over an inch deep. The seed should be early planted, or as soon as vegetation will proceed rapidly; but it must first be soaked, by pouring soft, scalding water on it, allowing it to cool to blood heat, and remain for one or two days, then roll in plaster and drill it in. The husk of the seed is thick and scarcely pervious to moisture, and without previous thorough saturation the seed will not readily germinate.

The culture is similar to that of carrots and parsneps. They should be thinned to a distance of about eight inches, and all vacancies filled with strong thrifty plants. It is better to sow thick enough to avoid the necessity of transplanting, for in addition to the time and expense of this operation, the new plants will not thrive as well as those which grow in their ranks from the seed. The above distances are suit able for the sugar beet. The mangold-wurzel attains a larger size, and the spaces may be increased. The practice of plucking off the leaves for cattle-feeding, is objectionable, as it materially interferes with the growth of the plants. Scarcely any disease or enemy troubles it, except when young. It is then sometimes, though rarely, attacked by grubs or small insects.

Harvesting may be commenced soon after the first leaves turn yellow, and before the frosts have injured them. The tops must not be too closely trimmed, nor the crown of the roots or its fibrous prongs cut from such as are destined for late keeping. If intended for early winter use, they may be abridged a trifle, and after the surface is dry, stored like other roots They do not need as effectual protection as potatoes; for if the frost reaches them under a covering of earth, it will gradually withdraw on the approach of warm weather, and leave the roots uninjured; but they will not keep as long as if untouched by the frost. A slight opening for the escape of the gas, as with other roots, should be left at the top, and partially guarded with straw.

Uses.—The beet is a universal favorite for the table, and of great value for stock. Domestic animals never tire of it, and swine prefer it to any other root excepting the parsnep. I have kept a large herd in the best condition through the winter, on no other food than the raw sugar beet. They possess additional merit, from their capability of resisting decay longer than the turnep, and frequently beyond the carrot and parsnep. They will be solid, fresh and juicy, late in the spring, if properly stored; and at a time too when they are most wanted for ailing sheep or cattle, milch cows or ewes, or for contributing to the support and health of any of the farm stock.

When fed to fattening animals, they should follow, and never precede the turnep. It has been found, that animals continue steadily to advance in flesh, after being carried to a certain point with turneps, if shifted on to the beet; but in repeated instances, they have fallen back, if changed from beets to turneps.

Davy found in 1,000 parts, the following quantity of nutritive or soluble matter.—White or English turnips, 42; Swede, 64; mangold-wurzel, 136; sugar beet, 146. This order of nutritive quality is followed by Boussingault, though he places the field beet and Swede turnep, at nearly the same point. Einhof and Thaer, on the contrary, place the Swede before mangold-wurzel. But in feeding to animals, unless for an occasional change, the roots should be given out in the order named.

The sugar beet is seen to be more nutritious than the mangold-wurzel; it is equally hardy and productive, and more palatable to stock, and of course is to be preferred as a farm crop. The former has been largely cultivated in France and Germany, for making into sugar, where it has been entirely successful, because protected by an adequate impost on the imported article. Their conversion into sugar, has repeatedly been attempted in this country, but it cannot sustain a successful competition with the sugar cane.

From the experiments of Darracq, it has been found that

in summer, the beet yielded from 3½ to 4 per cent. of sugar; but in October, after the commencement of frost, it gave only syrup and saltpetre, and no crystalizable sugar. When used for this purpose, the residuum of the pulp, after expressing the juice, is given to cattle. When wilted, the leaves are also fed to them, but caution, and the use of dry food in connection, is necessary to prevent their scouring. What are not thus used, are plowed in for manure. The beet is also distilled and yields about half the product of potatoes.

NOTE.

On the following pages, I append *the table of nutritive equivalents of food*, compiled by Boussingault, as a convenient reference, though not entirely reliable in all cases. For it will be seen, from what has before been said, that the particular plants, vary, not only according to the season and soil, but frequently also, according to the particular variety, subject to analysis. He says : " In the following table, to the numbers assigned by the theory, I have added those of the whole, which I find in the entire series of observations that have come to my knowledge. I have also given the standard quantity of water, and the quantity of azote, contained in each species of food. When the theoretical equivalents do not differ too widely from those supplied by direct observation, I believe that they ought to be preferred. The details of my experiments, and the precautions needful in entering on, and carrying them through, must have satisfied every one of the difficulties attending their conduct ; yet all allow how little these have been attentively contemplated, and what slender measures of precaution against error have been taken. In my opinion, direct observation or experiment is indispensable, but mainly, solely as a means of checking, within rather wide limits, the results of chemical analysis."

TABLE OF THE NUTRITIVE EQUIVALENTS OF DIFFERENT KINDS OF FORAGE.

TITLE.	Standard, water pr ct. prepared	Azote, pr ct.	Azote pr ct. not dried in the article	Theory	Block	Petit	Meyer	Thaer	Pabst	Flottow	Pohl	Rieder	Gemerhausen	Crud	Weber	Dombasle	Krantz	Schwertz	Schnee	Midleton	Murre	Andre	Boussin-gault.
Ordinary natural meadow hay	11-0	1-34	1-15	100	100	100	100	100	100	100	100	100	100	100	100	100	100	100	100	100	100	100	100
Ditto, of fine quality	14-0	1-50	1-30	98																			
Ditto, select	13-3	2-10	2-00	58	108	100				100		90	90	90	90	90	90	100	90				
Ditto, freed from woody stems	14-0	2-44	1-38	85				90	100					90						90		90	
Lucern hay	16-6	1-66	1-54	75	100	90		90	100	500	450	500	500						500				
Red clover-hay, 2d y'r's growth	10-1	1-70	0-64	311	430	90	156	450	425	175													
Red clover, cut in fl'r green do	26-0	0-36	0-27	426	200	360	156	450	300				660						666				
New wheat straw, crop 1841	76-0			235						175													
Old wheat straw	8-5	0-53	0-49	200	200	500	150	660	35				190					100	182				
Do. do. lower parts of stalk	5-3	0-43	0-41	280						175			160					100	154				
Do. do. upper part of do. and ear	9-4	1-42	1-33	86						175									143				
New rye straw	18-7	0-30	0-24	479	200	200	150	190	200	200			90						191				
Old ditto	12-6	0-60	0-42	250	230	180	150	150	200		90				150								
Oat-straw	21-0	0-36	0-25	383	193	200	150	180	160														
Barley ditto	14-0	0-30		460	165	250																	
Pea ditto	19-0	0-96	1-79	64					150					90									
Millet ditto	11-6	0-51	0-78	147		200		130	100					600					50				
Buckwheat ditto	9-2	1-18	0-48	240	160	200																	
Lentil ditto	11-0	1-01		114		125			600														
Vetches cut in flower and dried	11-0	1-16	1-14	101	600	300				600													
Potato tops	78-9	2-30	0-50	209				600															
Field-beet leaves	88-9	4-40	0-85	230																			
Carrot ditto	70-9	2-94	0-37	135				325	250														
Jerusalem potato stems	86-4	2-70	0-86	311	67		250	429	450														
Canada-poplar shoots	62-5	2-29	0-92	134	58			300	250														
Oak ditto	57-4	2-16	0-28	125	556			536															
Drum cabbage	92-3	3-70	0-17	411	500	500	290	460				600	600	500	600		400		600				
Swedish Turnep	91-0	1-83	0-13	676	300	300	250				330	370	350	525	500		400	200	350			350	
Turnep	92-5	1-70	0-21	885	533	600				500	525	555	525	255			200	450	525	800	667		
Field beet (1838)	87-8	1-70	0-21	548	366	400				300			460			261		333	460				
Ditto white Silesian	85-6	1-46	0-18	669	366											220							400

	Composition	Ratio	205	250	225	300	250	266	270	266	260	266	307	266	270	338	280	360	280
Carrots	87·6 / 2·40	0·30	389															360	280
Jerusalem potatoes (1839)	79·2 / 1·60	0·33	348																250
Ditto (1836)	75·5 / 2·20	0·42	274																
Potatoes (1838)	65·9 / 1·60	0·36	319	216	150	200	200	200		210	200	187		200				250		
Ditto (1836)	79·4 / 1·80	0·37	311	400															
Do. after keeping in the pit	76·8 / 1·18	0·30	383											227						
Cider apple pulp dr'd in the air	6·4 / 0·63	0·59	195															59		
Beet-root magma from sug. mill	70·0	0·38	303																	
Vetches in seed	14·6 / 5·13	4·37	26	54	50	56	40	50												
Field-beans	7·9 / 5·50	5·11	23	54		73	40	47												
White peas (dry)	8·6 / 4·20	3·84	27	54	49	66	40													
Kidney beans	5·0 / 4·30	4·58	25	39																
Lentils	9·0 / 4·40	4·00	29															40		50
New maize	18·0 / 2·00	1·64	70	52		76	50		52											
Buckwheat	12·5 / 2·40	2·10	55	64	53								47							
Barley (1836)	13·2 / 2·02	1·76	65	61		86	60	50	55											
Barley-meal	13·0 / 2·46	2·14	54																	
Ditto	13·0 / 2·20	1·90	61																	
Ditto	20·8 / 2·20	1·74	68	71	51	71	50	38	52					65						
Oats (1838)	12·4 / 2·22	1·92	61																	
Ditto (1836)	14·0 / 1·93	1·93	60	55	46	64	40						44	44						
Ditto (Parisian)	11·5 / 1·70	1·70	68	52													2			
Rye (1836)	11·5 / 2·27	1·50	77	55																
Ditto (1838)	10·5 / 2·33	2·00	58	52																
Wheat (1836, Alsace)	14·5 / 2·30	2·09	55	27																
Ditto (1838)	16·6 / 3·18	2·00	57																	
Ditto from highly manured soil	37·1 / 2·18	2·65	43					175												
Recent bran	7·6 / 0·94	1·36	85	105																
Wheat husks or chaff	13·4 / 1·39	0·85	135																	
Rice (Piedmont)	13·4 / 6·00	1·20	160																	
Linseed cake	5·0 / 4·78	5·20	96	42																
Hemp ditto	6·2 / 3·53	4·00	22																	
Beech mast ditto		4·21	27																	
Dry acorns	48·2 / 2·31	0·80	35																	
Refuse of the wine press air d'd		1·71	68	62		75														

THE JERUSALEM ARTICHOKE (Helianthus tuberosus).

This plant is a native of Brazil, but it has hitherto been little cultivated in this country. Loudon says the name Jerusalem is a corruption of the Italian word *girasole*, (sunflower), the blossom of which it closely resembles, except in size. It flourishes in a moist, loose soil or sandy loam, with little care except to thin out the plants and prevent weeds. It is hardy, very productive and easily cultivated in drills, three or four feet apart. The planting may be done in March or April. As it is not injured by frost and is prolific, it will spread rapidly and often becomes a nuisance in the garden. The product is enormous, sometimes overrunning 2,000 bushels per acre. Its nutritive qualities are much less than those of the potato ; but its greater productiveness and the facility of raising it, would seem to entitle it to more general favor. Boussingault considers it an improving and profitable crop, from its drawing its nitrogen largely from the atmosphere. It is peculiarly fitted for spring food, as the roots lie uninjured by the vicissitudes of the weather, and may be taken out in perfection after most other roots are gone.

The artichoke is used in this country, both for human and animal food. The roots are generally eaten as a pickle or salad. Loudon says, "they may also be eaten boiled, mashed in butter, or baked in pies, and have an excellent flavor." The tops when cut and cured as hay, afford a good fodder for cattle, and the roots are excellent for sheep and other stock. Swine will thrive upon them through the winter, and do their own harvesting when the ground is not locked up by frost.

CHAPTER X.

~~~~~~~~~~~~~~~~~~~

MISCELLANEOUS OBJECTS OF CULTIVATION

BROOM CORN (Sorghum saccharatum).

So far as I am acquainted with its history, this is a product peculiar to America. In its early growth and general appearance, it resembles Indian corn. It grows to the height of ten or twelve feet, with a perfectly upright stalk, from which an occasional leaf appears; and at the top, a long compact bunch of slender, graceful stems is thrown out, familiarly termed the *brush*, which sustain the seed at and near their extremities.

Soil.—The best soil for broom corn, is similar to that required for Indian corn or maize. It should be rich, warm, loamy land, not liable to early or late frosts. Spring frosts injure broom corn more than maize, as the roots do not strike so deep, nor has it the power of recovering from the effects of frost equally with the latter. The best crops are usually raised on a green sward, turned over as late as possible in the fall, to kill the worms. Clay lands are not suitable for it.

Manure.—Horse or sheep manure is the best for this plant ; and if mixed with much straw or other vegetables, they should be well rotted before applying. If the land is in good condition, three cords or eight to ten loads to the acre, is enough for one dressing. This is usually placed in hills, and 12 to 15 bushels of ashes per acre may be added with great advantage. Plaster, at the rate of two to four bushels per acre, is also beneficial. The addition of slacked lime helps the ground, affords some food to the crop, and is destructive to worms. Poudrette, at the rate of half a pint to each hill at planting, or Peruvian guano at the rate of a table-spoonful mixed in a compost with ten times its quantity of good soil, is an excellent application, especially if the land is not already rich enough. Repeating the above quantities around the stalks in each hill, after the last hoeing, will add materially to the crop.

Planting.—Broom corn should be planted in hills two feet apart, in rows two and a half to three feet distant. If the seed is good, drop 15 to 20 seeds in a hill, an inch and a half deep, to ensure eight or ten thrifty plants, which are all that must be left after the second hoeing. The time of planting must depend on the climate and season. The 1st of May is about the time in latitude 40°, and 10th to 15th in 42°; but let it be as early as possible, yet late enough to escape spring frost. Mellow the ground well with a harrow before putting in the seed. Thick planting gives the finest, toughest brush.

After Culture.—As soon as the plants are visible, run a cultivator between the rows, and follow with a hand hoe. Many neglect this till the weeds get a start, which is highly prejudicial to the crop. The cultivator or light plow should be used afterwards, followed with a hoe, and this may be repeated four or five times with advantage. Break the tops before fully ripe, or when the seed is a little past the milk; or if frost appears, then immediately after it. This is done by bending down the tops of two rows towards each other, for the convenience of cutting afterwards. They should be broken some 14 inches below the brush, and allowed to hang till fully ripe, when they may be cut and carried under cover, and spread till entirely dry. The stalks remaining on the ground, may be cut close, or pulled up and buried in the furrows for manure, and thus be restored to the earth to enrich it; or they may be carried to the yard to mix in a compost, with the droppings of the cattle.

Cleaning the Brush.—Unless some larger machine is used, this is most rapidly accomplished by passing it through a kind of hetchel, made by setting upright knives near together, or it may be cleaned by a long-toothed currycomb. By the use of the hetchel, none of the little branches are broken, and the brush makes a finer, better broom. We have seen horse power machines used for cleaning the seed with great rapidity, in the Miami Valley. The average yield is about 500 lbs. of brush per acre. It varies according to season or soil, from 300 to 1,000 lbs. The price also varies materially, ranging from three to sixteen cents per pound; the last being seldom obtained unless in extreme scarcity. A good crop of seed is matured in the Connecticut Valley, about two years out of five. When well ripened, the seed will average three or four pounds for every pound of the brush. A single acre has produced 150 bushels of seed, though 25 to 50 is a more common yield. It weighs about 50 lbs. per bushel, and is usually sold at 25 to 35 cents.

The uses of Broom Corn are limited to the manufacture of brooms from the brush; and the consumption of the seed when ground and mixed with other grain, in feeding to fattening or working cattle, sheep and swine, and occasionally to horses. Brooms manufactured from it, have superseded every other kind for general use in the United States, and within a few years, they have become an article of extensive export to England and other countries. The brush and wood for the handles are imported separately, to avoid high duties, and are there put together, and form a profitable branch of agricultural commerce to those engaged in the traffic. The cultivation of broom corn has, till quite recently, been almost exclusively confined to the northeastern States; but it is now largely raised in the Western. Their fresh, rich soil, however, does not yield so fine and tough a brush as that grown in the longer cultivated fields.

FLAX (Linum usitatissimum, Fig. 43).

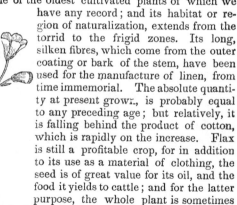

This is one of the oldest cultivated plants of which we have any record; and its habitat or region of naturalization, extends from the torrid to the frigid zones. Its long, silken fibres, which come from the outer coating or bark of the stem, have been used for the manufacture of linen, from time immemorial. The absolute quantity at present grown, is probably equal to any preceding age; but relatively, it is falling behind the product of cotton, which is rapidly on the increase. Flax is still a profitable crop, for in addition to its use as a material of clothing, the seed is of great value for its oil, and the food it yields to cattle; and for the latter purpose, the whole plant is sometimes fed with decided advantage.

Fig. 43.

The proper soil for Flax, is a good alluvial or vegetable loam, equally removed from a loose sand or tenacious clay. In a very rich soil, the fibres grow too coarse, and on a poor soil, the crop will not make a profitable return. Fresh barnyard manures are not suited to it, and they should, in all cases where necessary for a proper fertility, be added to the preceding crop. A rich sod which has long lain in pasture or meadow, well plowed and rotted, is the best for it. Lime in small quantities, may be incorporated with the soil, but the Flemings who raise flax extensive.y, never allow it to follow a heavy liming till seven years intervene, as they consider it injures the fibre. A good wheat, is generally a good flax soil. Salt, ashes and gypsum are proper manures for it, The last has the greatest effect, if applied after the plant is developed and while covered with dew or moisture. All the saline manures used as a top-dressing, benefit the plant and check the ravages of worms, which frequently attack the young plants.

Cultivation.—On a finely-prepared surface, either of fresh sod or after corn or roots, which have been well manured and kept clear of weeds, sow broadcast, from sixteen to thirty quarts per acre, if wanted for seed, or two bushels, if wanted for the fibre. When thin, it branches very much, and every sucker or offshoot is terminated by a boll well loaded with seed. When thickly sown, the stem grows single, and without branches, and gives a long, fine fibre. If the soil be very rich, and fibre is the object of cultivation, it may be sown at the rate of three bushels per acre. There is a great difference in seed; the heaviest is the best, and it should be of a bright brownish cast and oily to the touch. It must be lightly harrowed or brushed in and rolled. When three or four inches high, it may be carefully weeded by hand, and for this, it is best to employ children; or if adults are put on the field, they should be barefoot. Any depression of the plants by the feet will soon be recovered by the subsequent growth, which, on good soil, will be sufficiently rapid to prevent the weeds again interfering with it Grass seed or clover may be sown with flax without any detriment to it, but the flax ought to be too thick on the field to allow them to grow well; and in pulling, the flax, they will be very likely to be uprooted.

Harvesting.—When designed for cambrics and the finest linen, flax is pulled in flower; but in this country it is seldom harvested for the fibre till the seed is entirely formed, and although not ripe, most of it will mature on the stalk without any prejudice to the fibre. If required for seed, it must be left standing till the first s eds are well ripened. It is then gathered and bound in small bundles, and when properly dried, is placed under cover. If it falls before ripening, it should be pulled at once, whatever its stage of growth, as this is the only means of saving it.

After Management.—After removing the seed by drawing the heads through a comb or rake of finely-set teeth, called rippling, the usual method of preparing flax in this country, is by dew-rotting, or spreading it thinly on a clean sward, and turning it occasionally till properly prepared, after which it is put into bundles and stored till a convenient period for cleaning it. This is a wasteful practice and gives an inferior quality of fibre.

Water-rotting is the best plan of preparing it, which is done in vats or small ponds of soft water, similar to those used for hemp. This gives a strong, smooth, silky fibre, and without waste, and worth much more either for sale or for manufacturing, than the dew-rotted. Various steeps for macerating, and machines for preparing it have been used, which materially increase its marketable value; but it is generally got out on the *brake* by hand, when the farmer is most at leisure. A crop of the fibre may be estimated at 300 to 1,000 lbs.; and of seed, from 15 to 30 bushels per acre.

There are no varieties worthy of particular notice for ordinary cultivation. Great benefit is found to result from a frequent change of seed, to soils and situations differing from those where it has been raised. The seed is always valuable for the linseed-oil it yields, and the residuum of the seed or oil-cake, stands deservedly high as a feed for all animals. The entire seed when boiled, is among the most fattening substances which the farmer can use for animal food. Flax, like most other plants grown for seed, is an exhausting crop, but is not when pulled or harvested before the seed matures. The Flemings think flax ought not to be raised on the same soil oftener than once in eight years.

As a means of promoting the industrial interests of the country, the raising of flax, like that of hemp, cotton, wool and raw-silk, is an object of national importance. This value does not, like wheat, pork, butter, &c., end with their preparation for market, but constitutes a basis for other industrial occupations, after leaving the hands of the farmer. Each should be produced to the extent, at least, of supplying our own manufactures with the raw material, for making the fullest amount of fabrics we can consume at home or profitably export abroad.

HEMP (Cannabis sativa).

Large portions of our western soils and climate, are pecu-

FIG. 43.

liarly adapted to the production of hemp, and for many years it has been a conspicuous object of agricultural attention. We have not yet brought the supply to our full consumption; as we have till recently, imported several millions annually, either in its raw state, or as cordage, twine, sail-duck, osnaburgs and other manufactured articles. But the increased attention and skill bestowed on its cultivation, combined with our means for its indefinite production, will doubtless soon constitute us one of the largest of the hemp-exporting countries.

The Soil for hemp may be similar to that of flax, but with a much wider range, from a uniform standard; for it will thrive in moderately tenacious clay, if fertile, and well pulverized ; and it will do equally well on reclaimed muck beds, when properly treated. New land is not suited to it till after two or three years of cultivation. A grass sod or clover bed is best for it, when plowed in the fall or early in winter. This secures fine pulverization by frost and the destruction of insects, and especially the cut-worm, which is very injurious to it. If not already sufficiently mellow, it should be re-plowed in the spring, as a deep, fertile tilth is essential to its full vigor and large growth.

Cultivation.—Early sowing produces the best crop, yet it should not be put in so early as to be exposed to severe frost ; and where there is a large quantity planted, convenience in harvesting requires that it should ripen at different periods. The farmer may select his time for sowing, according to his latitude, and the quantity cultivated. From the 10th of April to 10th June is the fullest range allowed. The choice of seed is material, as it 's important to have a full set of plants on the ground ; yet an excess is injurious, as a part are necessarily smothered after absorbing the strength of the soil, and they are besides, in the way of the harvesting, without contributing anything to the value of the crop. Seed of the last year's growth is best. as it generally heats by being kept over, which can be avoided only by spreading thin From four to six pecks per acre of good seed is sufficient. The best is indicated by its weight and bright reddish color. It is usual to sow broadcast, harrow in lightly both ways, and roll it. A smooth surface is material in facilitating the cutting. Sowing in drills would require less seed, give an equal amount of crop, and materially expedite the planting. This should always be done before moist weather if possible, as rapid and uniform germination of the seed is thus more certainly secured. If the soil be very dry, it is better to place the seed deeper in the ground, to reach a proper moisture, which can be done with the plow. If sown in drills and well covered, it might be previously soaked so as to secure early germination in the absence of rains. I quote from an excellent article on hemp raising, in the American Agriculturist, by the editor, A. B. Allen.

Cutting.—" No after cultivation is necessary. When the blossoms turn a little yellow, and begin to drop their leaves, which usually happens from three to three and a half months after sowing, it is time to cut the hemp; if it stands a week or ten days longer than this, no other detriment will ensue except that it will not rot so evenly, and becomes more laborious to break. Cutting is now almost universally practiced in preference to pulling. Not quite so much lint is saved, but the labor is easier and all subsequent operations, such as spreading, stacking and rotting. The lint also is of a better color and finer fibre, and the roots and stubble left in the ground and plowed under, tend to lighten the soil, and are equivalent to a light dressing of manure. If the hemp is not above seven feet high, it can be cut with large and strong cradles, at the rate of an acre or two per day; but if above this height, strong brush scythes must be used, about two and a half feet long.

Drying and Securing.—As fast as cut, spread the hemp on the ground, taking care to keep the butts even, when if the weather be dry and warm, it will be cured in three days. As soon as dry, commence binding into sheaves, and if destined for water rotting, it ought to be transported to dry ground convenient to the pools, and then secured in round stacks, carefully thatched on the top to keep out the rain; but if designed for dew rotting, it should be secured in large ricks, in the same field where grown. The reason why these are to be preferred is, that less of the hemp is thus exposed to the weather, and the more and better the lint when rotted and broken out.

9

The Ricks should be 30 to 40 feet long, and 15 to 20 feet wide, the best foundation for which is logs laid down for the bottom course, six feet from each other, then lay across these, rails or poles one foot apart. As the hemp is bound in sheaves, let it be thrown into two rows, with sufficient space for a wagon to pass between. While the process of taking up and binding is going on, a wagon and three hands, two to pitch and one to load, is engaged in hauling the hemp to the rick, and stacking it. Thus the process of taking up, binding, hauling, and ricking, all proceed together. In this way five hands will put up a large rick in two days and cover it. For making the roof of the rick, it is necessary to have long hemp, from which the leaves should be beaten off. In this state only will hemp make a secure roof.

In laying down the hemp, begin with the top ends of the bundle inside, and if they do not fill up fast enough to keep the inside of the rick level, add, as occasion may require, whole bundles. Give it a rounded form at each end, and, as it rises, it must be widened, so as to make the top courses shelter the bottom ones. After it is twelve feet high, commence for the roof, by laying the bundles crosswise, within a foot of the edges of the rick, building the top up roof-shaped, and of a slope at an angle of about forty-five degrees. For the covering of the roof lay up the bundles at right angles to its length, the butt ends down, and the first course resting on the rim of the rick as left, one foot in width. Lap the bundles in covering the roof in courses, as if shingling a house. Commence the second course by reversing the bundles, placing the top ends down, and then go on lapping them as before. Begin the third course of shingling with the butt ends down, letting the first hang at least one foot below the edge of the roof, to shed off the rain from the body of the stack. Unbind the bundles, and lay the covering at least one foot thick with the loose hemp, lapping well as before, and for a weather board, let the top course come up above the peak of the roof about three feet and be then bent over it, towards that point of the compass from which the wind blows least. If the work has been faithfully performed, the rick may be considered as finished, and weather proof, and it requires no further binding. The rick should be made when the weather is settled, for if rain falls upon it during the process, it will materially injure the hemp. There ought always to be a sufficient number of hands in the field to gather, bind the shocks, and finish the ricking in a single day.

Time of dew rotting.—The best time for spreading hemp for dew rotting, is in the month of December. ' It then receives what is called a winter rot, and makes the lint of the hemp a light color, and its quality better than if spread out early. But where a farmer has a large crop, it is desirable to have a part of his hemp ready to take up late in December, so that he may commence breaking in January. To accomplish this object, a part of his crop may be spread about the middle of October. It would not be prudent to spread earlier, as hemp will not obtain a good rot if spread when the weather is warm. The experienced hemp-grower is at no loss to tell when the hemp is sufficiently watered. A trial of a portion of it on the break will be the best test for those who have not had much experience When sufficiently watered, the stalks of the hemp lose that hard, *sticky* appearance or feel, which they retain till the process is completed. The lint also begins to separate from the stalk, and the fibres will show themselves, like the strings of a fiddle-bow, attached to the stalk at two distant points, and separate in the middle. This is a sure indication that the hemp has a good rot.

Shocking after breaking and rotting.—When the hemp is dry, put in shocks of suitable size, without binding. Tie all the shocks together with a hemp-band, by drawing the tops closely, to prevent the rain from wetting the inside. Each shock should be large enough to produce from fifty to sixty pounds of lint. If the hemp be considerably damp when taken up, leave the shocks untied at the tops until they have time to dry. If not well put up, they are liable to blow down by a strong wind. To guard against this, it is desirable, when commencing a shock, to tie a band around the first armful or two that may be set up and then raise up the parcel so tied, and beat it well against the ground so as to make it stand firmly, in a perpendicular direction. The balance of the shock should now be set regularly around the part as herein directed. If hemp be carefully shocked, it will receive little or no injury till the weather becomes warm. In the meantime it should be broke out as rapidly as possible. If the operation be completed by the middle of April, no material loss will be sustained. If delayed to a later period, loss of lint will be the consequence. Cool, frosty weather is much the best for hemp-breaking. If the hemp is good, first-rate hands on the the common hemp break, will clean two hundred pounds per day. The ordinary task for hands is one hundred pounds.' —*Beatty.*

Hemp-brake.—The hand hemp-break is made like that for flax, only much larger; the under slats on the hinder end are 16 to 18 inches apart, at the fore-end they approach within three inches of each other. The slats in the upper jaw are so placed as to break joints into the lower as it is brought down on to the hemp. After breaking out the hemp, it is twisted into bunches, and sent to the press-house to be bailed, and is then transported to market.

Water rotting.—The best plan for water rotting is in vats under cover, the water in which is kept at an equal temperature. The hemp thus gets a perfect rot at all seasons of the year, in seven or ten days, and when dried, is of a bright, greenish, flaxen color, and is considered by many, of a better quality than the finest Russian, and it brings as high a price in market. These vats may be easily constructed and managed, and if built in a central position, by a company of planters on joint account, they would be but of small expense to each, and all in turn could be accommodated by them. The hemp is first broken previous to rotting, in a machine, which is moved by steam power; this lessens the bulk greatly, by ridding it of most of its woody fibre; but the process is not essential to rotting in vats. If to be rotted in spring or river water, artificial pools or vats must be formed for this purpose, which should not be over three feet deep, otherwise the hemp is liable to an unequal rot. It will require plank placed upon it weighted down with timbers or stones, in order to keep it well under water. Mr. Myerle recommends vats 40 feet long, 20 feet wide, and two feet deep. The hemp is thus kept cleaner while rotting, and the hands can lay it down in the vats and take it out without getting wet, which is important to the health of the laborer. These vats greatly facilitate the operation, and they can be fed with water and have it run off at pleasure, without loss from the hemp. Water rotting in streams, requires a longer or shorter period, according to the season. In September, when the water is warm, ten days is generally sufficient; in October, about fifteen, and in December, thirty days or more. For the latitude of Kentucky, October and November are considered the best months for the operation; and it is then easiest done, gives more lint, and as good a sample as if deferred later."

Raising Hemp Seed.—This requires another system of

cultivation, but on a similar soil, which should be in the finest condition as to fertility and pulverization. An old pasture or meadow heavily manured and plowed in the fall, and well pulverized in the spring, furnishes the best soil. I quote from Judge Beatty's essay on practical agriculture :

" The seed should be planted either in hills or drills. I prefer the former, because it admits of easier and better cultivation, as the plow can be used both ways. It is usual to plant five feet apart, each way, and suffer four or five stalks to stand in a hill until the blossom hemp is removed, and then reduce the number so as not to exceed two stalks in a hill. Thus there would be two seed plants for each twenty-five square feet. It would be a better practice to make the hills three feet six inches apart, each way, and thin the hemp to three stalks in a hill, till the blossom hemp appears ; at the proper time, cut out the blossom or male hemp; and if necessary a part of the seed hemp, so as to reduce the latter to one stalk in the hill.

The ground for hemp seed, having been well pulverized by plowing twice, and running the harrow, lay off as above directed, and plant in the same manner as corn. Twelve or fifteen seed should be scattered in each hill. Soon after the hemp comes up, run a small plow both ways, once in a row. If the ground is not foul, the plowing may be delayed till the hemp is a few inches high, which will enable the plowman to avoid throwing the dirt on the tender plants. The hoes should follow the second plowing, and clean away the weeds, in or near the hill, and thin out the hemp to seven or eight stalks. These should be the most thrifty plants, and somewhat separated from each other. Repeat the plowing to keep the ground light and free from weeds. When the plants are about a foot high, the hoes should again go over the ground and carefully cut down any weeds or grass which may have escaped the plow. The plants should be still further thinned out at this time, leaving but four in a hill, and some fine mold drawn around the plants, so as to cover any small weeds that may have come up around them. After seed hemp has attained the height of a foot and a half, it will soon be too large to plow, but it ought to have one plowing after the last hoeing. The ground, by this time, will have become so much shaded by the hemp plants, as to prevent the weeds from growing, and nothing more need to be done but for a man to follow the plow, and

if three and a half feet be the distance of the hills apart, reduce the number of plants invariably to three, taking care
to remove those which the last plowing may have broken.

When the seed hemp has so far advanced as readily to
distinguish the male from the female plants, let all the blossom hemp be cut out, except one stalk in every other hill,
each way. This will leave one stalk of male hemp for every
four hills. These, together with the stalks which afterwards blossom, will be sufficient to fertilize all the seed-
bearing plants, and secure a crop of perfect seed. After the
blossom plants have remained until they have discharged
their pollen, which can be easily ascertained by dust ceasing
to flow from them when agitated, they, also, should be cut
down. Some top the seed plants when five or six feet high,
to make them branch more freely, but this is not necessary
where but one or two seed-bearing plants are suffered to remain in each hill."

A seed-bearing hemp crop is a great exhauster of land,
while such as is grown only for the fibre, takes but a moderate amount of fertilizing matter from the soil. Unlike most
crops sown broadcast, it grows with such strength and luxuriance, as to keep the weeds completely smothered; and it
may, therefore, be grown for many successive seasons on the
same field, without the latter becoming foul. Its entire
monopoly of the ground, prevents the growth of clover or
the grasses in connection with it.

The seed yields an oil of inferior value, but when cooked,
it affords a fattening food for animals.

COTTON (Gossypium, FIG. 44).

Within the last few years, this has become the leading
agricultural export of the United States. The total amount
of the cotton crop in this country, in 1845, was estimated at
about 850,000,000 lbs., and for 1848, nearly 1,000,000,000
This enormous product has mainly grown up within the
last 60 years. Even as late as 1825, our total production
was within 170,000,000. The introduction of Whitney's
cotton gin, in the latter part of the last century, gave the first
decided movement towards the growth of American cotton.
Previous to this, the separation of the cotton seed from the
fibre was mostly done by hand ; and the process was so slow
and expensive, as to prevent any successful competition with
the foreign article. This incomparable invention, which
cleaned 1,000 lbs. in the same time a few pounds could be

cleaned without it, overcame the only obstacle to complete success; and millions of acres of the fertile lands of the South and West are now annually covered with the snowy product. The increase seems to know no check or abatement; as with the lessening price and increasing quantity, the demand seems constantly to augment.

Climate and Soil.—Cotton will grow in some of the middle States, but with little profit north of the Carolinas and Tennessee. The soil required is a dry, rich loam. Light sands and tolerably heavy clays will produce cotton, but with much less profit than the loams.

Fig 44.

Planting.—During the winter, the land intended for planting should be thrown up in beds, by turning several furrows together. These beds may be four feet from centre to centre for a moderate quality of upland soil, and five feet for the lowlands. But these distances should be increased with the increasing strength of the soil, to seven and eight feet, and in some instances, even to a greater distance for the strongest lands. These may lie until the time of planting, from 20th of March to 20th of April, when no further danger from frost is apprehended ; then harrow, and w th a light plow, mark the centre of the beds and sow at the rate of two to five bushels per acre. A drilling machine might be made to answer this purpose better, and save much time. An excess of seed is necessary, to provide for the enemies of the plant and other contingencies. If all the seed germinates, there will be a large surplus of plants, which must be removed by thinning. There is an advantage in mixing the seed before it is sown, with moistened ashes or gypsrm, as it facilitates sowing and germination. It should be buried about an inch deep, and the earth pressed closely over it.

The cotton plant, in the opinion of Dr. Philips, is hardier, even, than corn, when properly treated. He contends, and I must say, with a great deal of apparent reason, that thick planting is to be deprecated, and that but two or three pecks of prime seed should be sown to the acre. This would expose each plant to the atmosphere, and give it opportunity and room to develop a strong stalk with abundant foliage ; and he asserts, when thus grown, a frost that will cut down corn, will leave the young cotton plant uninjured. There is a great disadvantage in exposing the long, spindling shoots, that have grown up among a crowd of others, to the sun, wind and frost, when they are thinned out. A temporary check is thus always given to the plant, and frequently, a permanent injury is sustained in consequence of it.

The varieties of seed used for uplands are the Mexican and Petit-Gulf, both of similar origin, but the latter is better selected and has been kept pure. New seed which has been brought from a distance, ought to be substituted for the old as often as once in three years. The beneficial effects of this practice would probably be enhanced by bringing it from a different soil, and more especially, such as had grown upon one of inferior quality to that intended to be planted.

The Petit-Gulf, raised near Rodney, in Mississippi, is universally esteemed the best seed, and it will frequently yield from one to two hundred pounds more of cotton per acre than the average of such as is produced elsewhere This difference has arisen, in part at least, from the great care used in the selection and management of the seed and it may be, and probably is, in part due to some peculiar difference in soil or locality, that better prepares the seed for an abundant yield when transferred to remote fields.

The cultivation is performed with various instruments, the bull-tongue or scooter the shovel, double shovel, the sweep the harrow, the cultivator and the hoe. One or more of the former must be used to pulverize the land and uproot and clean off the weeds ; while the last is necessary to carry this operation directly up to the stem of the plants. The culture is thus summarily stated by Dr. M. W. Philips, of Mississippi ; " Commence cleaning the cotton early, and clean it well ; return to it as soon as possible, throw earth or mold to the young plants, and if the ground be hard give it a thorough plowing ; keep the earth light and mellow and the plants clear of grass and weeds." The plants are thinned at every hoeing, till they attain a height of three or four inches, when they are allowed to stand at intervals of about eight or ten inches for a medium quality of soil. This distance should be largely increased when it is richer.

Enemies and Diseases.—Cotton is subject to the cut-worm, the army-worm, and boll-worm, the slug, the caterpillar, cotton lice, grasshoppers, rot, sore shin and rust. I have seen no remedies prescribed for either, but suggest for experiment, the exposure of the two former to frost, by plowing just before its appearance. Late planting, and when the season is so well advanced as to give a rapid and uniform growth, is one of the most effectual preventives against disease or injury from insect enemies. The free use of lime and salt, and similar manures might arrest or mitigate the effects of all. Birds should also be encouraged upon the fields, as they would destroy numbers of the worm and insect tribes. It has been claimed that the introduction of the Mexican and Petit-Gulf varieties, is the most effectual remedy, as they are hardier kinds, are less the object of attack and have a greater ability to withstand it.

Harvesting is commenced when the bolls have begun to expand, and the cotton is protruded. This is continued as the bolls successively ripen and burst their capsules. It is done entirely by hand, the picker passing between two rows and gleaning from each. The cotton is placed in a bag capable of containing fifteen or twenty pounds, which is hung upon his shoulders or strapped upon his breast.—These are emptied into large baskets which are taken, when filled, to the gin-house. I quote the above authority : " Having all things ready for picking cotton, I commence early, and as soon as the hands can gather even twenty pounds each. This is advisable, not only in saving a portion of that from being destroyed, if rains should fall, which often do at this season, (about the middle of August), but for another reason ,

9*

passing through the cotton has a tendency to open out to sun and air, the limbs that have interlocked across the rows, and hastens the early opening. On low grounds, especially, much loss is incurred in some seasons from the want of the sun to cause an expansion of the fibre within the boll, so as to cause it to open. The boll is composed of five divisions, in each of which there is a parcel of cotton wool surround- ing each seed, there being several in each *lock* of cotton. When green, these fibres lie close to the seed, and as it ri pens, the fibres become elastic, the boll becoming hard and brownish. The Sea Island has only three divisions, as also the Egyptian, which is only the Sea Island of the best va- riety, with black seed, smooth, and a yellowish tuft of fibres on the small end ; they are both from Pernambuco. Some of the cotton we plant has only four divisions, but I think five gene ally.

There is a peculiar art in gathering the cotton from the boll, which can only be acquired by practice; many gather equally fast with either hand. The left hand seizes the stem near the open boll, or the boll between the two middle fin- gers, the palm of the hand up ; the fingers of the right hand are inserted tolerably low down in the boll, a finger on each lock of cotton ; then, as the fingers grasp it, there is a slight twisting motion, and a quick pull, which, if done well, will extract the contents.

Cotton should be gathered from the field as clean as possible, taken to the scaffolds and dried until the seed will crack when pressed between the teeth, not crush or mash, but crack with some noise. It should be frequently turned over and stirred, and all the trash and rotten pods taken out, while this is done, to insure its drying earlier.

If seeds are wanted for planting, gin the cotton imme- diately, and spread the seed over the floor some five inches thick, until perfectly dry. If the cotton-seed be not wanted, pack the seed-cotton away in the house, to remain until a gentle heat is discovered, or until sufficient for ginning. After it has become heated, until there is a sensible feeling of warmth to the hand, and it looks as if pressed together, open and scatter to cool. This cotton will gin faster, have a softer feel, is not so brittle, therefore not so liable to break by rapidity of gin, and has a creamy color ; the wool has imbibed a part of the oil that has exuded by the warmth of seed, and is in fact restored to the original color. I have known of a number of sales made of this description of cotton, and even those who are most strenuous against the heating, admit it bore a better price." The cotton is then ginned and baled, when it is ready for market.

Topping Cotton between the 20th July and 20th August is practiced by many planters with decided success. It is generally considered highly beneficial in dry seasons, but not in wet, and that in three years out of five it is attended with particular advantage to the crop.

SEA-ISLAND COTTON.

This crop is raised on the islands and low lands that border the coast of South Carolina and Georgia. If removed from this locality, the fibre seems immediately to degenerate. It requires, in many respects, a treatment unlike that of the upland. I insert an article by Thomas Spalding, Esq., an experienced and intelligent planter, who has long been engaged in its cultivation: " The Sea-Island cotton was introduced into Georgia from the Bahamas. The seed was from a small island near St. Domingo, known as Ar guilla, then producing the bestcotton of the western world It in no way resembles the Brazil cotton, which is the kidney-seed kind, introduced some years later, and which after trial, was rejected in Georgia. It came in small par cels from the Bahamas, in the winter of 1785 ; and gradually made its way along the coast of Georgia, and passed into Carolina. The winter of 1786 in Georgia was a mild one, and although the plants of the Sea-Island that year had no ripened their seed, it being a perennial, and subject only to be killed by frost, it started the next season from the roots of the previous year, its seed ripened, and the plants became acclimated. Many changes have come over this seed since that time, from difference of soil, culture, and local position, and above all, from careful selection of seed. But the cause is yet to be discovered, why the gain in fineness of wool, is lost in the quality and weight of the product ; for in spite of a zeal and intelligence brought to act upon the subject without parallel, the crops are yearly diminishing, until to grow Sea-Island cotton is one of the most profitless pursuits within the limits of the United States.

Planting.—When the first seed was introduced, it was planted in hills prepared upon the level field, at five feet each way ; but it was soon found to be a very tender plant, liable to suffer by storms, by wind, by drought, and by ex cess of rain. The quantity of seed was therefore increased, and the plants multiplied. If the seed is at first covered more than two inches, it will not feel the influence of the sun, and will not vegetate later in the season; that is, in April. You must give from three to four inches of covering to preserve the moisture, or you fail from an opposite cause, the wind and sun drying the soil too much for vegetation. In most countries, after sowing the seed, the roller is applied; but in cotton-planting, in our ridge-husbandry, the foot in covering the seed and pressing down the earth, well supplies its place.

Preparing the land.—Early in February, clean the rested fields, and either burn off the fennel-weeds and grass of the previous year, or list them in at five feet apart, to serve as the base of the future ridges or bed. There is much difference of opinion upon the subject of burning or listing in. I am inclined to take the first opinion, believing that the light dressing of ashes the field receives from burning off, is more beneficial to the soil than the decay of the vegetable matter, and renders it less liable to produce what is a growing evil, the rust; a species of blight, much resembling the rust or blight upon wheat, and which takes place about the same period, just as the plant is putting out, and preparing to ripen its fruit. For many years it has been the practice, among experienced planters, to divide the enclosed fields into two portions; the one at rest, the other in culture.

Ridging.—The land being listed in short lines across the entire field, at five feet apart, the operation of ridging is commenced about the first of March. The ridges occupy the entire surface, the foot of one ridge commencing where the other ends, and rising about eight inches above the natural level of the land, thus presenting a surface almost as smooth, and almost as deeply worked, as a garden-bed. This is done but a few days ahead of the planting, and the ridge is from two to two and a half feet broad at top; it is then trenched on the upper surface with the hoe, six inches wide, and from three to six inches deep, depending upon the period of planting.

Quantity of seed per acre.—A bushel of seed is generally sown to the acre, I believe half a bushel is better; for where the evil comes, whether the worm, wind, drought, or wet, there is no security in the many, but where they come up thin, they soon grow out of the way of injury.

After-culture.—The cultivation of Sea-Island cotton is carried on by the hand-hoe, and the quantity is always limited to four acres to the laborer. The operation of weeding commences as soon as we finish planting, because in our flat and sandy soils the grass-seed springs with the first growth of the cotton, and by the time we finish planting, say the 1st of May, what we planted in March requires the hoe. In the operation of hoeing and weeding, the land is kept, as far as may be, at its original level, the beds neither increased nor diminished, that the heavy rains which generally fall in August, may injure the growing plants, which are then in full bearing, as little as possible. The young cotton is thinned out slowly at from six to twelve inches apart on the ridge, by the 10th of June. As soon as the rains commence, which is about the last of July, it is wise to leave nature to herself, and no longer disturb the soil. Four hoeings, if well done, and the grass well picked at each hoeing, is enough; nor does any aftergrowth of grass do injury.

Manures and soiling stock.—For ten years past, great efforts have been made by the Sea-Island planters in manuring. Much of the alluvion of our salt rivers has been collected, and sometimes placed directly in heaps through the fields at rest, and at other times placed in cattle-pens on which cotton seed and all waste materials are strewn, and the cattle penned upon it. But what is preferred, is to pen our cattle near the river at night, and cut salt-grass, which covers these alluvion lands, and which is as nutritious as so much clover. Great benefits will hereafter, undoubtedly result from the use of marl.

Amount of crop per acre and picking.—It has been stated already, that 500 pounds to the acre, is about the medium crop, which at 20 cents per pound (more than the actual price for the last three years), is to the planter, $100 for gross crop; and from this $100 dollars, is to be subtracted bagging, freight, expenses of sale, clothing for his people, medical attention, and too often provisions."

The varieties of Cotton in the United States, which have been cultivated with success in addition to those enumerated, are *the Rio,* with a staple about three inches in length of a glossy, silky texture, brought from South America; *the Egyptian,* received from the garden of Mehemet Ali, and grown in Louisiana 15 feet in height; *the Mastadon,* lately introduced from Mexico, firm in texture and quite productive; *the Chinese Silk Cotton,* white, soft, fine and silky; *the East India,* growing to a height of 14 feet, and producing a beatiful fibre; and *the Nankeen,* a handsome staple of a true nankeen color, raised by the late Hon. John For syth, of Georgia, and some other planters.

Cotton seed, like most cf our cultivated plants, is the creature of circumstances, and improves or deteriorates within certain limits, according to climate, soil, manner of selecting, &c. Even the long, silken fibre, and the black seeds of the Sea-Island, assume the shorter, coarser staple, and the green seed of the upland, when exposed for two or three years to the same soil and position; and a corresponding change takes place,when the upland is made to grow on the low sea-girt islands of South Carolina and Georgia. The sea-muck, which is full of minute shells (lime), decomposed animal matter, including many of the phosphates and salt, is one of the best applications for improving the quality of staple, as well as augmenting the quantity of fibre. Salt, by many experienced planters, is deemed the principal ingredient in effecting this improvement. It is believed that it may be applied with equal success to the upland varieties.

The amount of seed in cotton is large, being nearly 70 per cent. of the entire gathering, the fibre being about 28. This is used for various purposes. Sometimes it is pressed for the oil, of which it yields from 15 to 20 per cent. of its own weight. When thus treated, the cake is used for cattle food. The seed is frequently, though improperly fed raw to stock; and this often proves fatal, especially to swine when fed before the adhering fibre has been decomposed, besides being attended with much waste. It is most advantageously prepared by boiling for half an hour, when it will benefit all descriptions of stock. By adding an equal quantity of corn, and boiling them together, it will fatten swine rapidly. It is also useful to land as a manure.

SUGAR CANE (Saccharum officinarum, Fig. 45).

This plant was cultivated in Louisiana, on the present site of the city of New Orleans, as early as 1726. For more than half a century, however, its use was almost wholly limited to the conversion of its juice into syrup and molasses. Sugar was not made within the State of Louisiana, till after 1760. In 796, the second sugar mill was erected on the plantation now occupied by the town of Carrolton. Owing to the revolution in St. Domingo, and the emigration of large numbers of her planters to Louisiana. the cultivation of the cane was largely extended soon after this period.

yet so late as 1818, the entire crop of the State amounted only to 25,000 hogsheads.

Steam power was first applied to grinding the cane, in 1822, and since that period, its cultivation has been rapidly spreading over the delta of the Mississippi, and the adjoining territory. The product of Louisiana in 1845, reached the enormous quantity of 207,337,000 pounds, and about 9,000,000 gallons of molasses, worth nearly $15,000,000 ; being an increase of over ten times the quantity yielded but 30 years before.

Fig. 45.

Its introduction into other States.—The cane was introduced into Georgia from Otaheite, in 1805, and sugar was for sometime produced for export. It was an object of attention while it commanded ten cents per pound ; but when the prices declined to five and six cents, it ceased to be manufactured as an article of commerce, though still extensively produced for domestic consumption. It has recently engrossed the attention of planters to some extent, in Florida and Texas ; and there are portions of these States eminently calculated, both in soil and climate, to rival the best sections of Louisiana in its culture. The southern portions of Alabama and Mississippi, as well as the Carolinas, have also begun to plant the cane for the supply of their own syrup, and they have thus far proved it an object well worthy their pursuit.

The introduction of the Ribbon Cane, from Georgia into the adjoining States, in 1817, by giving a much hardier variety, has largely extended the area of its cultivation, and rendered it a desirable crop, where it would not otherwise have been an object of attention. It is probable, too, that it is gradually becoming acclimated, and that hereafter, it will endure an exposure, and yield profitable returns far north of any point where it has hitherto been successfully grown.

The total product of sugar has not yet reached over 60 per cent. of our own consumption. But this interest is rapidly extending throughout those portions of the southern States adapted to it; and with the immense area, capable of being reclaimed from the swamps of the Gulf coast and elsewhere, devoted to this object, there is no improbability that the energies of American planters, will soon carry our production beyond the total of the world, in 1844, which was below 780,000 tons, or about 1,500,000 hogsheads.

Varieties.—The kind most cultivated in the United States, is *the striped Ribbon or Java,* which is by far the hardiest and most enduring cane. It grows rapidly, is of large size, and resists the effects of early and late frosts, and the excess of rains or drought and disease, better than any other. It has, however, a hard, coarse rind, and yields juice of only a medium quality. The outer coating is beautifully striped with alternate blue and yellow, of varying widths, and changing in every successive joint. *The red ribbon, or violet,* from Java, is much like the foregoing, except in having a uniform color, and by many it is preferred for new land.

The Creole, crystaline or Malabar, was the first introduced, and though of diminutive size, is a cane of great richness and value. Several varieties of *the Otaheite, the purple, the yellow, and the purple-banded,* are more or less cultivated. Some of these were brought into Georgia at an early day, and thence transferred to Louisiana. Those I have seen have a large stalk of great succulency, but yielding a juice decidedly inferior to the Creole. Some additional varieties, such as *the grey canes,* intermixed with the ribbon, and occasionally others have been introduced, but they are not of superior quality, or of general cultivation.

Soil.—The cane will flourish in a great variety of soils, varying between the extremes of a stiff clay and a light sandy loam, provided the former be well drained and fertile. The soil best suited to it is a fertile loam, well supplied with 'ime, and such as will yield the best crops of Indian corn. Some of the best and most enduring soils in the West Indies and elsewhere, contain large quantities of lime and the phosphates. The most profitable sugar plantation in Louisiana, has a profusion of shells scattered over it, in every stage of decomposition.

Seed Cane.—This plant is always propagated by cuttings. These ought to be provided from the best cane of the preceding season. From the use of the unripe tops and close, negligent planting, it is supposed the Creole cane has degenerated to its present diminutive size. There is less vigor and growth usually from tops than from the ripened cane; though where these are matured and a portion of the ripe stalk is left with it, the resulting crops, under the most favorable circumstances, are scarcely distinguishable. Fine growth frequently follows planting the tops, where the land is new and fertile ; but it is the exception and not the rule ; and good husbandry dictates a reliance on sound, mature cane only, for general use. There is no doubt, that in accordance with the general laws of vegetation, the cane crop would be benefitted by a change of the plant cane from one section of country to another. But as this would be attended with so much trouble and expense as to preclude the undertaking, the only remaining means available for securing improvement or preventing deterioration, is to be found in planting the best qualities of seed cane.

Preservation of the Cane.—This is kept from the period of cutting till planting, by simply placing it on the dry surface of the field, in beds or *mattresses,* as they are technically termed, of about two feet in depth, and having the tops shingling or overlying the ripe portion of the stalk. The tops should lie towards the south, to prevent their being lifted and frozen from severe north winds, which sometimes occur. Thick beds preserve a more uniform temperature, and repel the approach both of frost and the sun's rays ; thus serving the double purpose, besides their preservation from frost, of preventing fermentation during fall and winter, and germination on the approach of spring.

It is well to preserve an excess of seed cane, as continued and severe spring frosts may cut down and destroy so many young shoots, as to leave a deficiency, unless partially replanted. Many assert the cane will keep better by being cut soon after a rain, so as to be bedded with the sap vessels full, and that dry rot follows when cut after a long drought. Some, however, allow it to lie on the ground and wilt for two or three days after cutting, and think when thus treated, it keeps equally well.

Cane which is intended for grinding is often thus secured, when severe frosts are anticipated. It requires additional labor to top and trim it when thus harvested ; but a good yield of sugar is in this way often secured, which might otherwise be lost.

Preparation for planting.—Where the land is new (as much of it is, that is now appropriated to cane), it is invariably light and full of vegetable matter. Shallow plowing,

and wide distances between the rows, are here justified
The cane grows luxuriantly in such soils, and where there
is a deficiency of warm weather to mature it fully, as in
Louisiana, room is requred to allow a free circulation of
air, and the full benefit of the sun, to ripen it before the approach of frosts. From seven to ten feet is near enough for
the rows, but these should contain from two to three continuous lines of good plant cane. Where the land is fertile,
wide rows, if well cultivated, will produce an equal quantity as if planted closer, and there is much less expense and
labor in planting and tending the crop.

Land that has been long in cultivation, may be planted
nearer; but if sufficiently fertile, as it ought always to be,
it should never be nearer than six feet, and under certain
circumstances, may extend to nine. It was formerly the
practice to plant a single line of seed cane, in rows from two
and a half to four feet apart; but this system has been given
up, as it was found troublesome in cultivating, slower in
ripening, and it is believed materially and permanently to
have lessened the size of the cane first introduced.

Some planters make their cane beds every sixteen feet,
planting in each, two rows at a distance of four feet, and
leaving a space between every alternate row of eleven feet.

There is a great advantage in these wide spaces, as the
trash (tops, leaves, and all dead vegetable matter left on the
ground), and *bagasse* (*megasse*, [Fr.] the residuum of the
cane after expressing the juice), can all be buried between
the widest spaces, and remain undisturbed till decomposed,
without prejudice to the growing crop. On light or sandy
lands, these materials may be burned and the ashes applied
to the soil; but in adhesive or clay lands, good husbandry
requires that all this should be buried, as the vegetable decay (carbonaceous matter or humus), not only contains every
element for the reproduction of the future crop, but it effects
a mechanical division in the soil, of great value to its porosity, friability and productiveness. Occasionally, the trash
is buried at the foot of the plant, in which situation the
earth is kept constantly upon it. Some place the cane at a
distance of ten or twelve feet, and plant corn between the
rows, which matures and withers before the cane reaches
its full size. Others sow the cow pea, while still occupied
with cane, to renovate the land; but neither practice seems
to meet with general favor, as they interfere with the main
purpose of planting, which is to produce the greatest quan-

tity of mature cane, an object that can only be secured by its thorough and exclusive cultivation on the field.

The land should be deeply broken up with a two or four horse plow. If light or sandy, it may be plowed flat; but if stiff, or too much inclined to wet, it should be thrown into beds. Great advantages have generally followed the use of the subsoil plow, when run a foot below the bottom of the turning furrow, and immediately under the rows to be occupied by the cane. This is the more important, as no opportunity will again occur for breaking up this portion of the field, till the plant is renewed. The plowing may be done at any convenient time between October and March; but on plantations where the harvesting and grinding are going forward, it is seldom the plows can be started before the last of December, or early in January. The plowing should never long precede the planting, unless in stiff soils, which need the meliorating influence of the atmosphere to crumble the massive clods; nor in these, beyond the period necessary to effect this object. A fine bed of well-pulverized earth is thus secured for the plants to root in, and afford its nourishment to the young shoots.

Planting.—This may be done any time between October and April. There is a greater certainty of a good crop if in the ground by the first of March. The occupations of harvesting, grinding, plowing, &c., will usually postpone the commencement of planting till January. On the land previously well plowed, open a wide furrow with the fluke or double mold-board plow. Clean this out with the hoe of a uniform width, by the removal of any clods that may have fallen in after the plow. With the increased width now usually adopted by the best planters, not less than three parallel seed stalks should be planted. These ought to be precisely in line, and at least four inches apart; and it is better to place them so that the eyes may shoot out horizontally, and thus come up at the same time and on opposite sides of the stalk. Cover with sufficient earth to prevent freezing from any weather that may follow. On the approach of spring, remove the earth to the depth of one or two inches. Light spring frosts will not otherwise injure the cane, than to cut down the young shoots, and thus delay the growth till new leaves appear. The danger is in removing so much of the earth as to expose the roots to freezing.

Cultivating.—Throughout most of Louisiana, the cane

yields three crops from one planting. The first season it is called *plant cane* and subsequently, *ratoons*. In the tropical climates, the West Indies and elsewhere, ratoons will frequently continue to yield profusely for twelve or fifteen years. On new and peculiarly favorable spots in this country, the ratoons will produce equal to the plant cane for several years, occasionally for six or eight; and sometimes, as on the prairies of Attakapas and Opelousas, and the higher northern range of its cultivation, it requires to be replanted every year. The cultivation is alike in each, after the young shoots make their appearance; previous to which, the ratoons should be *barred* off and *scraped* on the approach of settled warm weather. The former consists in running the plow near the rows and throwing the earth from them; and the latter, in removing the soil from the surface. The sun's influence is thus sooner felt upon the roots, and an earlier and more prolonged growth is secured to the cane. But if these operations are performed too early or too closely, subsequent frosts may seriously injure the plants. Scraping has in some instances been partially accomplished by a large and cumbersome machine, but thus far it seems not to have been generally adopted. One could probably be constructed for the accomplishment of both purposes, that would save much labor and produce a uniformly beneficial result.

Soon after the young plants have made their appearance, the earth is gradually thrown to them by repeated plowings, and the hoes are made auxiliary to this object, and to keeping the cane clear of weeds. There is a great advantage in wide planting, as the two-horse plow can be used for cultivating. With these, a greater width and depth of furrow is secured, by which one plowman with two mules, will perform nearly double the amount of work, and do it much more thoroughly, than with the single plow. There is generally a larger growth from this deep and efficient plowing; and where weeds, and especially the *coco grass* abound, great economy in subduing these is secured by the use of the large plow, as it is thus so deeply buried, as to find its way to the surface only after long intervals.

When the cane has acquired such a height and expansion of leaves as to shade the ground effectually, which if all preliminary operations have been well performed, will be by the first of June, the last furrows are thrown to the roots and the earth slopes gradually to the centre; forming an elevated support to the plants, and a depression between, which serves as a drain for the surplus rains. Many intelligent planters run a large subsoil plow two or three times between the rows, which serves to loosen the soil for the greater extension of the roots, and this also more effectually drains the land. Throughout the operation of cultivating, after the ratoons have been barred off in the spring, care should be observed to avoid cutting or breaking the roots. This caution is applicable to all plants, but especially to the cane, which requires all the aid from its roots to mature before the approach of cold weather.

Deep plowing, both in breaking up and in cultivating, is essential to good sugar crops, on all lands that have long been subject to tillage. On the alluvial lands of the Mississippi, fresh upturned soil, exposed to the surface for the first time, always brings with it new supplies of food for the plants; and the more perfect and general the pulverization of the soil, the greater is the space afforded for a range to the roots. Good implements and good plowmen are essential preliminaries to a good sugar crop.

Harvesting.—In the West Indies and most other foreign States where grown, the cane fully ripens. This is true, also, with some of that on Tampa Bay, Florida, the cultivation of which has recently been considerably extended. But in Louisiana, the cane never fully matures. It begins to ripen at the foot of the stalks in August or September, and advances upwards at the rate of about six inches per week. The proper period for cutting, would be just previous to the heavy or black frosts (*freezes* they are generally called); but as it requires several weeks to secure the crop, much of which would be liable to great injury if left beyond the proper period, the harvesting is generally commenced by the middle of October, and steadily followed up till completed. This is done by striking off the top (unripened stalks), then stripping the leaves by a single downward stroke of the knife on either side, and another blow severs one or more stalks at the foot. The cane is then thrown into carts and hauled to the mill, where it should at once be ground, boiled, granulated and put up for market.

The moment of interference with the natural condition of the plant, is the signal for breaking up its normal or healthy condition, and sending its elements rapidly forward on a new career of change. The exquisitely-arranged crystals of sugar, which may be seen with the microscope, closely wedged in their appropriate cells within the silicious rind of the stalk, are susceptible to the slightest alteration in any portion of the plant. By cutting it at the bottom, the air gains access to the exposed cells, the sugar combines with the oxygen of the atmosphere and induces the first step towards decomposition, called the vinous fermentation, by which alcohol is developed. A second speedily follows, termed the acetic (the distinguishing peculiarity of vinegar), and this, if not arrested, soon terminates in the destructive or putrefactive fermentation, by which all the useful or nutritive properties of the cane are destroyed, and its materials are converted into their original elements, or are worthless for any purpose but manure. This change goes forward slowly with the sugar cane, while the temperature is low, but rapidly as it becomes elevated.

Slight frosts in autumn are beneficial rather than injurious, as by deadening the leaves and tops they check vegetation, and stimulate rather than retard the ripening of the plant. When severe frosts are apprehended, it will justify cutting the cane as rapidly as possible, and matrassing as before described, under the head of *the preservation of cane.* When thus shielded from the approach of the elements by the overlying cane, and at a period when the average temperature is near the freezing point, scarcely any change is perceptible for many weeks. The same result follows when remaining slightly frozen for an indefinite period. But the moment a thaw commences, the nitrogenized matter of the stalk mingles with the sugar, through the rupture caused by the expansion of the cells from frost, the oxygen of the air gains access, and fermentation begins, after which it is impossible to convert the saccharine matter into sugar. Molasses, alcohol and vinegar are the only forms which the crystaline matter of the cane can then be made to assume.

The amount of the products of cane, depends on several circumstances, the kinds planted, the soil, the season, manner of grinding, and the subsequent treatment in its conversion into sugar.

The quantity of the crop of sugar varies from 500 to 3,000 pounds per acre, the last amount only being realized under the most favorable circumstances. In good seasons, and with skillful treatment, 2,000 pounds are often obtained ; but owing to adverse causes, and negligent management, it is doubtful whether the average crop of the country comes up to 800 pounds per acre.

The composition of cane of a medium quality, is water,

72; woody fibre, 10; and sugar, 18; in every 100 parts; yet, notwithstanding the water and sugar (juice) constitute 90 per cent. of the cane, the best horizontal rollers have been made to express from 70 to 75 per cent. only, while the more imperfect grinding often reduces this below 50 per cent. It is estimated that the average product is about 56 per cent. of juice, leaving 34-90ths or more than one third of the entire quantity still in the bagasse, and wholly unavailable for any economical purpose. This shows a great deficiency in the mechanical operations of sugar-making; and it is the more to be regretted, as we know that the minute grains of sugar exist as perfect in the cane, as in any subsequent state of its granulation. Boussingault asserts, that he has seen the juice of the cane, under the skillful treatment of the chemist yield nothing but crystalizable sugar.

Value of the products of cane for animal food.—Large quantities of the molasses have heretofore been used for distilling into alcohol, but the manufacture of this has materially lessened of late, and a salutary change has been made in its disposal. When it would not bring a remunerating price for exportation, as has sometimes been the case in the West Indies, it has been mixed with other materials, and fed to stock. It is healthful and exceedingly fattening to animals. Its great value for conversion into fat will be readily seen, by comparing the elements of each. Sugar, which is identical with syrup and molasses, except that the two latter contain more water, and often some salts and other impurities, in suspension, has been analyzed by several chemists, with slightly varying results. According to the following authorities, it consists in every 100 parts, of

	Lussac & Thenard.	Berzelius.	Prout.	Ure.
Oxygen,	50.63	49.856	53.35	50.33
Carbon,	42.47	43.265	39.99	43.38
Hydrogen,	6.90	6.875	6.66	6.29

Fat, according to Chevreul, consists of 79 carbon; 11.4 hydrogen; and 9.6 of oxygen. Thus, it will be seen, that fat and molasses, are identical in their constituents, though varying in their relative proportions; and it would be fairly inferable from theory, as it has been found in practice, that no food is better suited to the easy and rapid conversion into animal fat, than sugar and molasses.

The process of Sugar-making, is one rather belonging to the arts, than to agriculture; and my limits will prevent

any description of it. Much attention has been devoted to
the subject within the last few years, and great improvements have been the result. But when fully carried out in
detail, so large an expenditure is required, as to preclude
their adoption by the mass of moderate planters. The apparatus of Messrs. Degrand, Derosne, Cail, Rillieux, and
others, including defecators, steam-jackets, Dumont filters,
vacuum pans, steam-pipes, and other improvements, may in
whole or in part be advantageous to the large planters, and
by many of these they have been adopted.

It is much to be desired, that the two objects of raising
the cane and converting it into sugar, could be separated,
like most other purely agricultural and mechanical operations. This change in the arrangement of sugar production, would effectually break up the aristocratic feature
which characterizes our present sugar estates, and which is
at such utter variance with nearly all the other branches of
our agricultural pursuits. Sugar estates might then be
divided among smaller proprietors, each of whom, by having a common market for his cane, would receive its full
value, whether it were one or one thousand acres. We
should thus witness an improvement in both the rearing of
the cane and its manufacture, which is not likely to be so
fully or speedily attained in any other way.

Manures for the cane.—If the alluvial bottoms of Louisiana and other fertile lands are properly managed, they will
never become exhausted by the cultivation of cane. *Tired
of it* they may be, as land is of any one constantly-recurring
crop; but *exhaustion* will never be realized, if the elements
constituting the stalk, and not converted into sugar, be returned to the soil. This is done, simply by burying the bagasse and trash. If the former is burned, as is sometimes
the case where there is deficiency of fuel, the ashes should
be carried to the field. The elements of the sugar, which
is the only portion necessarily or permanently withdrawn
from the field, are such as abound in the atmosphere, rains
and dews, and are profusely brought to it, by every passing
breeze and every falling shower. The inorganic or earthy
portions are, therefore, the essential constituents to be returned to it. To show the proportions of these, the analysis of
the ash of the cane is subjoined, as given by Mr. Stenhouse.

ANALYSIS OR THE ASHES OF THE SUGAR CANE.

	1	2	3	4	5	6	7	8	9	10	11	12
Silicia.........	45 97	42.90	46.46	41.37	46.48	50.00	45.13	17.64	26 38	52.20	48 73	54.59
Phosph'o acid	3 76	7.99	8.23	4.59	8.16	6.56	4.88	7.37	6 20	13 04	2 90	8.01
Sulphuric acid	6.66	10.94	4.65	10.93	7.52	6 40	7.74	7.97	6.03	3.31	5 35	1 93
Lime..........	9.16	13.20	8.91	9.11	5 78	5.09	4.49	2.34	5.87	10.64	11 62	14.36
Magnesia.. ...	3.66	9.88	4.50	6.92	15 61	13.01	11 90	3.93	5 48	5.63	5 61	5.30
Potassa	25.50	12.01	10 63	15.99	11.93	13 69	16.97	32.93	31.21	10.09	7.46	11.14
Soda..........	....	1.39	....	....	0.57	1.33	1.64	...	....	0.80	....	....
Chlo'. potass'm	3.27	....	7.41	8 96	....	....	....	10 70	11.14	....	16.06	0.84
Chlo'. sodium..	2.02	1.69	9 21	2.13	3.95	3.92	7 25	17.12	7.64	4.29	2.27	3.83

" Nos. 1, 2, 3, and 4, were very fine, full-grown canes, from Trinidad, consisting of stalks and leaves, but without the roots. Nos. 5, 6, and 7, were similar canes from Berbice ; No. 8, from Demarara ; No. 9, of full-grown canes, but with leaves, from the Island of Grenada; No. 10, from Trelawny, Jamaica, consisting of transparent canes in full blossom, grown about six miles from the sea, and manured with cattle dung ; No. 11, of transparent canes, from St. James', Jamaica, growing about two hundred yards from the sea, being old ratoons, and also manured with cattle dung ; No. 12, young, transparent canes, three and a half miles from the sea, and manured with cattle dung, guano and marl.—*Prof. Shepard.*"

Herapath found that 1,000 grains of the cane when burned, left but 7½ grains of ash, which was made up of inorganic bases, in nearly the following proportions, viz. : silicia, 1.8 ; phosphate of lime, 3.4 ; oxide of iron and clay, .2 ; carbonate of potash, 1.5 ; sulphate of potash, .15 ; carbonate of magnesia, .4 ; and sulphate of lime, .1.

The amount of fresh cut cane from an acre is sometimes enormous, exceding, probably, in some instances, 30 tons ; but where the trash and bagasse are restored to the soil, nothing more is required to sustain its fertility ; yet there may still be a failure of the crop from the neglect of rotation. Many throw out their land to accomplish this object by rest ; and while thus lying apparently idle, an important change is wrought in the soil by the action of a new class of vegetable roots, the weeds and such chance vegetation as may happen to occupy the field, which rapidly prepares it for its accustomed crop. But this end is attained more certainly, by a dense covering of such plants as may be best adapted to the purpose. Some alternate with corn, but this will be seen to violate a cardinal principle laid down under the head of *rotation*, as it approaches too nearly to the cane in its character. Corn may take its place in the fields but

10

with far less benefit, *regarded as a feature of rotation,* then if they were occupied l y the cow pea or some other plants, widely differing in their peculiarities from the cane.   When the pea has been on the ground for one or two years, and especially if the crop has been turned under, an immense growth of the cane has followed.

Where manures are sought for exhausted fields, the table of the ash of cane would indicate that potash, in some form, is highly essential, as well as lime, salt, the sulphates and phosphates.   These, and the other fertilizing materials, can generally be procured in adequate proportions from stable manures, if the latter are to be had; but where there is a deficiency of them, the land may be restored, by adding most or all of the following materials.

Potash is one of the leading manures required by the cane ; and this may be procured from various sources. Ashes will afford it with the most economy, and in the greatest abundance.   It is yielded by the slow decay of vegetable matters, and stable manures.   It is also procured from the decomposition of many species of rocks and stones.   *Lime, marl or ground shells.*   These are mostly pure carbonate, with sometimes a slight addition of the phosphate of lime.   Immense quantities of these exist in large deposites, throughout the lower delta of the Missis-sippi, and with such a tendency towards decomposition, as to be easily broken down by an efficient mill.   *Gypsum* (sulphate of lime) is an appropriate and economical manure. *Ground bones* (phosphate of lime); *salt* (chloride of so-dium), and *charcoal,* are all efficient manures for cane.

Drainage, deep, thorough under-drainage, is peculiarly necessary in preparing the sugar lands of Louisiana to yield their utmost burthen, and choicest quality of sugar cane. Drainage should not be limited to surface ditches.   It should embrace a systematic net-work of under-drains, with tiles deeply laid below the surface, and beyond the reach of the sub-soil plow, even when buried in he deepest depressions between the rows.   All the advantages enumerated under the head of draining (Chap. IV.) will apply here.   The cane on such thoroughly-drained lands, will commence growing earlier in the spring than on the undrained; it will grow faster during summer, it will continue growing longer in au-tumn, ripen earlier and mature a larger portion of the stalk, and yield a sounder, richer juice.   The expense and constant repair of surface drains will be saved ; the large proportion of the field now taken up by them and their banks, w'll be avoided; there will be no damage to the crop from excessive rains; no baking on the surface, or washing of the finer particles of the earth into the ditches. The land saved by this system would pay for carrying it out; and sometimes, even a single crop would fully repay it, which might otherwise be lost by long continued rains. The cane would always be better, and could by no possibility be worse than it now is. Where there are stiff lands, and the object could be achieved by no other means than by the disposal of one half the plantation, it is probable the annual net profit derivable from the remainder, when thus improved, would be greater than the whole without it. A system of under-drainage, would of course, necessarily imply the use of leading ditches and draining wheels, wherever adopted throughout the low-lands of the Mississippi Valley. Until this great desideratum can be accomplished, the most complete arrangement of surface drains should be fully carried out.

MAPLE SUGAR.

Fig. 46.

The Sugar, Rock, or Hard Maple Tree (*Acer saccharinum*, Fig. 46), is among our most beautiful shade, and most valuable forest trees; and it stands next to the sugar cane in the readiness and abundance with which it yields the materials for *cane sugar*. When refined, there is no difference either in appearance or quality, between the sugar from the cane, the maple or the beet. In the brown state, the condition in which it is sent to market, when made with care and formed into solid cakes, it retains its peculiar moisture and rich aromatic flavor, which makes it more acceptable to the nibblers of sweets, than the most refined and highly scented *bon-bons* of the confectioner. The quantity made in this country is very large; though from the fact of its domestic consumption, and its seldom reaching the principal markets, there is no estimate of the aggregate production which will come very near the truth. The product for Vermont alone, for 1845, was estimated at over 10,000,000 lbs. The quantity supposed to be annually sold in the city of New York exceeds 10,000 hhds. Both the sugar and syrup are used for every purpose for which the sugar from the cane is employed.

The sugar maple extends from the most northerly limits

of Maine and the shores of Lake Superior, to the banks of the Ohio. Farther south it is rarely found. The cane and maple approach each other, but scarcely meet ; and never intermingle as rivals in the peculiar region which nature has assigned to each. In some sections of the country, the sugar maple usurps almost the entire soil, standing side by side like thick ranks of corn, yet large and lofty, and among the noblest specimens of the forest. Immense quantities of these are to be found throughout the original forests of our northern, western, and middle States. I have seen them for miles in extent, near the borders of Lake Superior—a continuous wilderness of the sugar maple. I have also seen them in Wisconsin, near Lake Michigan, as they are found in the natural *sugar orchards* of that beautiful State. In these, they grow in open land among the rich native grasses, their tops graceful and bushy, like the cultivated tree ; and but for their greater numbers and extent, and their more picturesque grouping, one would think the hand of taste and civilization had directed, what nature alone has there accomplished. Amidst those beautiful orchards, or in the depths of those dense dark woods, the Indian wigwam and the settler's rude cabin may be seen, filled with the solid cakes and mo-koks,* each of which contain from 30 to 60 lbs., of their coarse-grained, luscious sugar.

The season for drawing and crystalizing the sap is in early spring, when the bright sunny days and clear frosty nights give it a full and rapid circulation. The larger trees should be selected, and tapped by an inch augur, to the depth of an inch and a half, the hole inclining downward to hold the sap. At the base of this, another should be made from three eighths to half an inch diameter, in which a tube of elder or sumach should be closely fitted to lead it off. A rude contrivance for catching the sap is by troughs, generally made of the easily-wrought poplar ; but it is better to use vessels which admit of thorough cleaning, and these may be suspended by a bail or handle from a peg driven into the tree above. When the sugar season is over, the holes ought to be closely plugged, and the head cut off

---

* Mo-kok—An Indian sack or basket, with flattish sides and rounded ends, similar in fashion to a lady's travelling satchel. They are made perfectly tight, from strips of white birch bark, sowed with thongs of elm. Many of the sap buckets are made of the same material, but different in form. The small mo-koks, tastefully ornamented with various-colored porcupine quills and filled with maple sugar, are sold as toys.

evenly with the bark, which thus soon grows over the wound. If carefully managed, several holes may be made in a thrifty tree without any apparent injury to it. The barbarous, slovenly mode of half girdling the trunk with an axe, soon destroys the tree.

The sap is collected daily with buckets, which are carried to the boilers on the neck, by a milk-man's yoke. If the quantity be great and remote from the sugar fires, a hogshead may be used for this purpose. This is placed on a sled, with a large hole at the top, covered with a cloth strainer, or a tunnel similarly guarded, is inserted in the bung ole. The primitive mode of arranging the sugary, is with large receiving troughs placed near or partially within the cabin, and capable of holding several hundred gallons of sap. The boiling kettles are suspended over the fires, on long poles supported by crotches.

The process of sugar making I give from the statement of Mr. Woodworth, of Watertown, N. Y., who obtained the premium from the State Agricultural Society, for the best sample of maple sugar, exhibited at the annual fair of 1844. The committee, who awarded the premium, say " they have never seen so fine a sample, either in the perfection of the granulation, or in the extent to which the refining process has been carried ; the whole coloring matter is extracted, and the peculiar flavor of maple sugar is completely eradicated, leaving the sugar fully equal to the *double* refined cane loaf sugar. The statement says : " In the first place, I make my buckets, tubs and kettles all perfectly clean. I boil the sap in a potash kettle, set in an arch in such a manner that the edge of the kettle is defended all around from the fire. This is continued through the day, taking care not to have anything in the kettle that will give color to the sap, and to keep it well skimmed. At night I leave fire enough under the kettle to boil the sap nearly, or quite to syrup, by the next morning. I then take it out of the kettle, and strain it through a flannel cloth into a tub, if it is sweet enough ; if not, I put it in a caldron kettle, which I have hung on a pole in such a manner that I can swing it on and off the fire at pleasure, and finish boiling, then strain it into the tub, and let it stand till the next morning. I then take this, and the syrup in the kettle, and put it altogether in the caldron, and sugar it off. To clarify 100 lbs. of sugar, I use the whites of five or six eggs, well beaten, about one quart of new milk, and a spoonful of saleratus, all well mixed

with syrup before it is scalding hot. I keep a moderate fire
directly under the caldron until the scum is all raised; then
skim it off clean, taking care not to let it boil so as to rise
in the kettle before I have done skimming it; when it is
sugared off, leaving it so damp that it will drain a little. I
let it remain in the kettle until it is well granulated; I then
put it into boxes, made smallest at the bottom, that will hold
from 50 to 70 lbs., having a thin piece of board fitted in, two
or three inches above the bottom, which is bored full of
small holes to let the molasses drain through, which I keep
drawn off by a tap through the bottom. I put on the top of
the sugar in the box, two or three thicknesses of clean, damp
cloth, and over that a board well fitted in, so as to exclude
the air from the sugar. After it has nearly done draining,
I dissolve it, and sugar it off again, going through the same
process in clarifying and draining as before.''

When sap is not immediately boiled, a small quantity of
lime water should be added to check fermentation, which
prevents the granulation of the syrup. A single tree has
yielded 24 gallons of sap in one day, making over seven
pounds of sugar; and in one season it made 33 lbs. Trees
will give an average of two to six pounds annually.

TOBACCO (Nicotiana, Fig. 47).

This narcotic is a native of
North America, and has been an
object of extensive use and cultivation in this country since the
first settlement of Virginia, in the
latter part of the 16th century. It
formed for a long time the principal export from that colony and
Maryland. It is still largely cultivated there, and has since become
an object of considerable attention
in the middle and western States,
and to some extent in the northern.

Fig. 47.   *The soil* may be a light, loamy
sand or alluvial earth, well drained and fertile. New land,
free from weeds and full of saline matters, is best for it;
and next to this, is a rich grass sod which has long remained
untilled. The seed should be sown in beds which must be
kept clean, as the plant is small and slow of growth in the
early stages of its existence, and easily smothered by weeds.

If not newly cleared, the beds ought to be burned with a heavy coating of brush.

Cultivation.—Pulverize the beds finely, and sow the seed at the rate of a table spoonful to every two square rods. The seeds are so minute, that sowing evenly is scarcely attainable, unless by first mixing with three or four times their bulk of fine mold. This should be done sufficiently early, to secure proper maturity to the plants in time for transplanting, (by the last of February or early in March south of the Ohio, and about the first of April north of it), covering lightly and completely rolling or treading down the earth. The plant appears in 15 or 20 days, and will be fit for transplanting in six or eight weeks. This should be done in damp weather, and the plants set singly, at a distance of two and a half to three feet each way. The after culture is like that of corn, and consists in frequently stirring the ground with the plow or cultivator and hoe, and keeping down weeds. The places of such plants as fail, or are blighted, should be at once filled up, and all worms destroyed.

The priming, topping, suckering and worming are necessary operations. The first consists in breaking off four or five of the leaves next the ground which are valueless; the second is taking off the top to prevent the seed stalk from developing, and is regulated by the kind of tobacco. "The first topping will always admit of a greater number of leaves being left; and in proportion as the season advances, fewer leaves should be left. The heavier kinds of tobacco are generally topped early in the season, to twelve leaves, then to ten, and still later to eight. The lighter kinds are topped to a greater number of leaves. If the soil is light, fewer leaves should be left." (*Beatty.*) Suckering consists in breaking off the young side shoots, which should be done immediately after they make their appearance. Worms of very large size and peculiarly destructive to the finer qualities of tobacco, abound during a part of the season. These can only be removed by repeatedly picking off by hand.

Harvesting may be commenced with such plants as nave matured, which is indicated by greenish yellow spots on the leaves. This will generally occur in August at the South, and in September at the North. The stem of the plant is cut near the surface and allowed to wilt on the ground, but not exposed to a hot sun. If there is danger of this, cut only in the morning or evening, and when properly wilted, which will be in a few hours, it may be carefully carried to the drying house, where it should be hung up by twine tied to the butt end of the stalk, and suspended over poles at drying distances with the head downwards. The circulation of air is necessary in the dry houses, but there must be entire safety against storms or winds, as the leaves are liable to break by agitation, and rain seriously injures them. When the stem in the leaf has become hard, it is sufficiently dried This takes place in good weather, in two or three months. The leaves may be stripped in damp weather when they will not crumble, and carefully bound in small bundles, termed hands, and then boxed for shipment.

The varieties of tobacco are numerous, not less than twelve being cultivated in America. They soon adapt themselves to the different soils and climates where they are grown. The most fragrant are produced in Cuba, and these are exclusively used for cigars. They command several times the price of ordinary kinds. The tobacco of Maryland and the adjoining States is peculiarly rich and high flavored, and is most esteemed for chewing.

Much of the peculiarity of taste and aroma, and the consequent value of tobacco, depends on the soil, and the preparation or sweating of the plant after drying. The former should not be too rich, and never highly manured, as the flavor is thereby materially injured, though the product will be increased. Yet it is an exhausting crop, as is seen by the large quantity and the analysis of the ash ; and the soil requires a constant renewal of well-fermented manures, and particularly the saline ingredients, to prevent exhaustion. Tobacco contains nitrogen and the alkalies in large quantities, and but very little of the phosphates. The ash is shown in the analysis of Fresenius and Will, to consist, of potash, 30.67 ; lime (mostly with a little magnesia), 33.36 ; gypsum, 5.60 ; common salt, 5.95 ; phosphates, 6.03 ; silica, 18.39 ; in 100 parts of the ash. The inferior kinds contain a large proportion of lime, and the superior qualities, the largest of potash.

The customary method of burning fuel on the beds designed for tobacco, and the use of freshly cleared and burnt lands, by which the largest crops of the best quality are obtained, shows conclusively the proper treatment required. By each of these operations, the ground is not only loosened in the best possible manner, and all insects and weeds de stroyed, but *the salts and especially potash in an available form,* are produced in the greatest abundance. Some ot the best soils in Virginia have been ruined by a constant succession of tobacco crops, the necessary result of neglect in supplying them with the constituents of fertility so largely abstracted. The yield per acre is generally from 1,500 to 2,500 lbs. It is a profitable crop when the best kinds are cultivated, under favorable circumstances of soil and climate. The total estimated product of the United States for 1843, was over 185,-000,000 lbs., of which Kentucky furnished 52,000,000, and Virginia nearly 42,000,000 lbs. Missouri, Ohio and other western States are rapidly becoming large producers.

INDIGO (Indigofera tinctoria, Fig. 48).

Indigo was formerly cultivated in the southern States, to a limited degree, but the introduction of cotton, the great profits which it yielded, and its consequent rapid extension, drove the culture almost entirely on to foreign soils. The decline in the price of cotton from large production, the increasing consumption of indigo in this country, together with the diminished price of other southern staples, will probably again make it an object of agricultural attention in those States where the soil and climate are suited to it. We have no detailed history of its cultivation in the United States, and I quote from Loudon.

FIG. 48.

He says, "it is one of the most profitable crops in Hindostan, because labor and land here are cheaper than any where else; and because the raising of the plant and its manufacture may be carried on without even the aid of a house. The first step in the culture of the plant is to render the ground, which should be friable and rich, perfectly free from weeds and dry, if naturally moist. The seeds are then sown in shallow dr lls about a foot apart. The rainy season must be chosen for sowing, otherwise if the seed is deposited in dry soil, it heats, corrupts, and is lost. The crop being kept clear of weeds is fit for cutting in two or three months, and this may be repeated in rainy seasons every six weeks. The plants must not be allowed to come into flower, as the leaves in that case become dry and hard and the indigo produced is

10*

of less val ie ; nor must they be cut in dry weather, as they would not sprir.g again. A crop generally lasts two years. Being cut, the herb is first steeped in a vat till it has become macerated, and has parted with its coloring matter ; then the liquor is let off into another, in which it undergoes the peculiar process of beating, to cause the fecula to separate from the water. The fecula is let off into a third vat, where it remains some time, and is then strained through cloth bags, and evaporated in shallow wooden boxes placed in the shade. Before it is perfectly dry it is cut in small pieces of an inch square ; and is then packed in barrels, or sowed up in sacks, for sale."

Indigo can only be raised to advantage in our most southern States. The soil requires to be dry, finely pulverized, and rich. The seed is sown early in April, in drills about eighteen inches apart, and the weeds are kept down by the hoe. It should be cut with the sickle or scythe, when the lower leaves begin to turn, and just before the plant is going into flower. This period occurs in this country, about the middle of summer. A second crop may be taken the first of autumn, and in hotter climates, even a third.

The consumption of indigo in this country already amounts to between two and three millions of dollars annually. *There are several varieties indigenous to the southern States,* and one or more in the northern, which yield inferior dye.

MADDER (Rubia tinctorum, Fig. 49).

F G. 4J.

The root of this plant is used for several dyes, but principally for the rich madder red ; and it has been recently an object of attention in the United States. The introduction of this, with numerous other articles consequent upon the extended growth of our manufactures, shows the intimate and mutually beneficial effects of associating the two leading industrial occupations of agriculture and manufactures The principal cause which has prevented its cultivation among us thus far, has been the long time required for maturing a crop. I subjoin a description of its culture from Mr Ba eham.

Soil and preparation.—" The soil should be a deep, rich, sandy loam, free from weeds, roots and stones. Alluvial bottom land is the most suitable, but it must not be wet. If old upland is used, it should receive a heavy coating of vegetable earth, from decayed wood and leaves. The land should be plowed very deep in the fall, and early in the spring apply about one hundred loads of well-rotted manure per acre, spread evenly and plowed in deeply, then harrow till quite fine and free from lumps. Next, plow the land into beds four feet wide, leaving alleys between three feet wide, then harrow the beds with a fine light harrow, or rake them by hand so as to leave them smooth and even with the alleys; they are then ready for planting.

Preparing sets and planting.—Madder sets or seed roots are best selected when the crop is dug in the fall. The horizontal uppermost roots with eyes are the kind to be used; these should be separated from the bottom roots and buried in sand, in a cellar or pit. If not done in the fall, the sets may be dug early in the spring, before they begin to sprout. They should be cut or broken into pieces, containing from two to five eyes each. The time for planting is as early in spring as the ground can be got in good order, and severe frosts are over, which in this climate is usually about the middle of April. With the beds prepared as directed, stretch a line lengthwise the bed, and with the corner of a hoe make a drill two inches deep along each edge and down the middle, so as to give three rows to each bed, about two feet apart. Into these drills drop the sets, ten inches apart, covering them two inches deep. Eight or ten bushels of sets are requisite for an acre.

After-culture.—As soon as the plants can be seen, the ground should be carefully hoed, so as to destroy the weeds and not injure the plants; and the hoeing and weeding must be repeated as often as weeds make their appearance. If any of the sets have failed to grow, the vacancies should be filled by taking up parts of the strongest roots and transplanting them; this is best done in June. As soon as the madder plants are ten or twelve inches high, the tops are to be bent down on to the surface of the ground, and all except the tip end, covered with earth shoveled from the middle of the alleys. Bend the shoots outward and inward, in every direction, so as to fill all the vacant space on the beds, and about one foot on each side. After the first time covering, repeat the weeding when necessary, and run a single norse plow through the alleys several times, to keep the earth clean and mellow. As soon as the plants again become ten or twelve inches high, bend down and cover them as before, repeating the operation as often as necessary, which is commonly three times the first season. The last time may be as late as September, or later if no frosts occur. By covering the tops in this manner, they change to roots, and the design is to fill the ground as full of roots as possible. When the vacant spaces are all full, there will be but little chance for weeds to grow; but all that appear must be pulled out.

The second year.—Keep the beds free from weeds; plow the alleys and cover the tops, as before directed, two or three times during the season. The alleys will now form deep and narrow ditches, and if it becomes difficult to obtain good earth for covering the tops, that operation may be omitted after the second time this season. Care should be taken when covering the tops, to keep the edges of the beds as high as the middle, otherwise the water from heavy showers will run off, and the crop suffer from drought.

The third year.—Very little labor or attention is required. The plants will now cover the whole ground. If any weeds are seen, they must be pulled out; otherwise their roots will cause trouble when harvesting the madder. The crop is sometimes dug the third year; and if the soil and cultivation have been good, and the seasons warm and favorable, the madder will be of good quality; but generally, it is much better in quality, and more in quantity, when left until the fourth year.

Digging and harvesting.—This should be done between the 20th of August and the 20th of September. Take a sharp shovel, and cut off and remove the tops with half an inch of the surface of the earth; then take a plow of the largest size, with a sharp coulter and a double team, and plow a furrow outward, beam-deep, around the edge of the bed; stir the earth with forks, and carefully pick out all the roots, removing the earth from the bottom of the furrow; then plow another furrow beam-deep, as before, and pick over and remove the earth in the same manner; thus proceeding until the whole is completed.

Washing and drying.—As soon as possible after digging, take the roots to some running stream or pump to be washed. Take large, round sieves, two and a half or three feet in diameter, with the wire about as fine as wheat sieves; or if these cannot be had, get screen-wire of the right fineness and make frames or boxes about two and a half feet long, and the width of the wire, on the bottom of which nail the wire. In these sieves or boxes, put half a bushel of roots at a time, and stir them about in the water, pulling the bunches apart so as to wash them clean ; then, having a platform at hand, spread the roots about two inches thick for drying in the sun. Carry the platforms to a convenient place, not far from the house, and place them side by side, in rows east and west, and with their ends north and south, leaving room to walk between the rows. Elevate the south ends of the platforms about eighteen inches, and the north ends about six inches from the ground, putting poles or sticks to support them—this will greatly facilitate drying. After the second or third day drying, the madder must be protected from the dews at night, and from rain, placing the platforms one upon another to a convenient height, and covering the uppermost one with boards. Spread them out again in the morning, or as soon as the danger is over. Five or six days of ordinarily fine weather will dry the madder sufficiently, when it may be put away till it is convenient to kiln dry and grind it.

Kiln drying.—The size and mode of constructing the kiln may be varied to suit circumstances. The following is a very cheap plan, and sufficient to dry one ton of roots at a time. Place four strong posts in the ground, twelve feet apart one way, and eighteen the other ; the front two four-teen feet high, and the others eighteen ; put girts across the bottom, middle and top; and nail boards perpendicularly on the outside, as for a common barn. The boards must be well-seasoned, and all cracks or holes should be plastered, or otherwise stopped up. Make a shed-roof of common boards. In the inside put upright standards about five feet apart, with cross-pieces, to support the scaffolding. The first cross-pieces to be four feet from the floor ; the next two feet higher, and so on to the top. On these cross-pieces, lay small poles about six feet long and two inches thick, four or five inches apart. On these scaffolds the madder is to be spread nine inches thick. A floor is laid at the bottom to keep all dry and clean. When the kiln is filled, take six or eight small kettles or hand furnaces, and place them four or five feet apart on the floor (first securing it from the fire with bricks or stones), and make fires in them with charcoal, being careful not to make any of the fires so large as to scorch the madder over them. A person must be in constant attendance to watch and replenish the fires. The heat will. ascend through the whole, and in ten or twelve hours it will all be sufficiently dried, which is known by its becoming brittle like pipe-stems.*

Breaking and grinding.—Immediately after being dried, the madder must be broken and ground immediately, or it will gather dampness so as to prevent its grinding freely. Any common grist-mill can grind madder properly; and when ground it is fit for use, and may be packed in barrels, like flour, for market."

Quantity per acre.—Mr. Swift, of Ohio, has raised 2,000 barrels per acre in one crop of four years growth, at a nett profit, including all charges of rent, labor, &c., of $200 per acre. The roots of madder are also a good food for cattle, but the expense and delay of producing it, will preclude its use for that purpose in this country.

WOAD (Isatis tinctoria, Fig. 50).

FIG. 50.

Woad is largely used in this country for dyeing, but generally, as a base for blues, blacks and some other colors, and for these it supplies the place of indigo. There are several varieties of woad, but the common biennial plant is the only one cultivated. Loudon says:—

" *The soil for woad* should be deep and perfectly fresh, such as those of the rich, mellow, loamy, and deep vegetable kind. Where this culture is carried to a considerable degree of perfection, the deep, rich, putrid, alluvial soils on the flat tracts extending upon the borders of the large rivers, are chiefly employed for the growth of this sort of crop; and it has been shown by repeated trials, that it answers most perfectly when they are broken up for it immediately from a state of sward.

---

* This seems to be a simple way of accomplishing the object, and within every one's reach; but as carbonic acid gas is thus constantly generated and closely confined, and by its gravity will occupy the lower strata of air, the greatest caution will be necessary for the person attending on the kiln drying, to prevent injury to himself

The preparation of the soil, when woad is to be grown on grass land, may either be effected by deep plowings with the aid of the winter's frost, cross plowing and harrowing in spring; by deep plowing and harrowing in spring; by paring and burning; or by trench-plowing, or spade-trenching. The first mode appears the worst, as it is next to impossible to reduce old turf in one year; and, even if this is done, the danger from the grub and wire-worm is a sufficient argument against it. By plowing deep in February, and soon afterwards sowing, the plants may germinate before the grub is able to rise to the surface; by trench-plowing, the same purpose will be better attained; and, best of all, by spade trenching. But a method equally effectual with the first, more expeditious, and more destructive to grubs, insects, and other vermin, which are apt to feed on the plants in their early growth, is that of paring and burning. This is, however, chiefly practiced where the sward is rough and abounds with rushes, sedge, and other plants of the coarse kind, but it might be had recourse to on others, with benefit.

The mode of sowing is generally broad-cast, but the plant might be most advantageously grown in rows, and cultivated with the horse hoe. The rows may be nine inches or a foot apart, and the seed deposited two inches deep. The quantity of seed for the broad-cast method is five or six pounds to the acre; for the drill mode, two pounds are more than sufficient, the seed being smaller than that of the turnep. New seed, where it can be procured, should always be sown in preference to old; but, when of the latter kind, it should be steeped for some time before it is put into the ground.—The time of sowing may be extended from February to July. Early sowing, however, is to be preferred, as in that case the plants come up stronger and afford more produce the first season. The after-culture of the woad consists in hoeing, thinning, prong-stirring, and weeding, which operations may be practiced by hand or horse tools.

Gathering the crops.—The leaves of the spring-sown plants will generally be ready towards the latter end of June or beginning of July, according to the nature of the soil, season and climate; the leaves of those put in at a later period in the summer are often fit to be gathered earlier. This business should, however, constantly be executed as soon as the leaves are fully grown, while they retain their perfec:

green color and are highly succulent; as when they are let remain till they begin to turn pale, much of their goodness is said to be expended, and they become less in quantity, and of an inferior quality for the purposes of the dyer. Where the lands are well managed they will often afford two or three gatherings, but the best cultivators seldom take more than two, which are sometimes mixed together in the manu-facturing. It is necessary that the after-croppings, when they are taken, should be constantly kept separate from the others, as they would injure the whole if blended, and con-siderably diminish the value of the produce. It is said that the best method, where a third cropping is either wholly or partially made, is to keep it separate, forming it into an inferior kind of woad. In the execution of this sort of busi-ness, a number of baskets are usually provided in proportion to the extent of the crop, and into these the leaves are thrown as they are taken from the plants. The leaves are detached from the plants, by grasping them firmly with the hand, and giving them a sort of a sudden twist. In favorable seasons, where the soils are rich, the plants will often rise to the height of eight or ten inches; but in other cicumstances, they seldom attain more than four or five.

The quantity produced is from a ton to a ton and a half of green leaves. The price varies considerably; but for woad of the prime quality, it is often from twenty-five to thirty pounds* the ton, and for that of an inferior quality, six or seven, and sometimes much more.

To prepare it for the dyer, it is bruised by machinery to express the watery part; it is afterwards formed into balls and fermented, re-ground, and fermented in vats, where it is evaporated into cakes in the manner of indigo. The haulm is burned for manure or spread over the straw-yard, to be fer-mented along with straw-dung. To save seed, leave some of the plants undenuded of their leaves the second year, and when it is ripe, in July or August, treat it like turnep seed. The only diseases to which the woad is liable, are the mildew and rust. When young, it is often attacked by the fly, and the ground is obliged to be re-sown, and this more than once, even on winter-plowed grass-lands.''

WELD OR DYER'S WEED (Reseda luteola, Fig. 51).

Weld is much used by the manufacturers of various

---

* The pound sterling may be reckoned at about five dollars

Fig. 51.

fabrics, as a dye. It has not to my knowledge been cultivated in this country. I again quote from Loudon : " Weld is an imperfect biennial, with small fusiform roots, and a leafy stem from one to three feet in height. It is a native of Britain, flowers in June and July, and ripens its seeds in August and September. Its culture may be considered the same as that of woad, only being a smaller plant it is not thinned out to so great a distance. It has this advantage for the farmer over all other coloring plants, that it only requires to be taken up and dried, when it is fit for the dyer. It is an exhausting crop.

Weld will grow on any soil, but fertile loams produce the best crops. The soil being brought to a fine tilth, the seed is sown in April or the beginning of May, generally broadcast.

The quantity of seed used, is from two quarts to a gallon per acre, and it should either be fresh, or if two or three years old, steeped a few days in water previously to being sown. Being a biennial, and no advantage obtained from it the first year, it is sometimes sown with grain crops in the manner of clover, which when the soil is in a very rich state, may answer, provided that hoeing, weeding and stirring take place as soon as the grain crop is cut. The best crops will obviously be the result of drilling and cultivating the crop alone.

Sowing.—The drills may be a foot asunder, and the plants thinned to six inches in the row. In the broadcast mode, it is usual to thin them to six or eight inches' distance every way. Often, when weld succeeds grain crops, it is never either thinned, weeded, or hoed, but left to itself till the plants are in full blossom.

The crop is taken by pulling up the entire plant ; and the proper period for this purpose is when the bloom has been produced the whole length of the stems, and the plants are just beginning to turn of a light or yellowish color; as in the beginning or middle of July in the second year. The plants are usually from one foot to two feet and a half in height. It is thought by some, advantageous to pull it rathei early, withou; waiting for the ripening of the seeds; as by this means there will not only be the greatest proportion of dye, but the land will be left at liberty for the reception of a crop of wheat or turneps; in this case, a small part must be left solely for the purpose of seed In the execution of the work, the plants are drawn up by the roots in small hand-fuls; and after each handful had been tied up with one of the stalks, they are set up in fours in an erec⁺ position, and left to dry. Sometimes, however, they become sufficiently dry by turning, without being set up. When dry, which is effected in the course of a week or two, they are bound up into larger bundles, each containing sixty handfuls, and weighing fifty-six pounds. Sixty of these bundles constitute a load, and in places where this kind of crop is much grown, are tied up by a string made for this purpose.

The produce of weld depends much on the nature of the season; and from half a load to a load and a half per acre is the quantity most commonly afforded. It is usually sold to the dyers at from five or six to ten or twelve pounds ($25 to $60) the load, and sometimes at considerably more. It is sometimes gathered green and treated like woad or indigo; but in general the dried herb is used by the dyers in a state of decoction.

The use of weld in dying is for giving a yellow color to cotton, woollen, mohair, silk and linen. Blue cloths are dipped in a decoction of it, which renders them green; and the yellow color of the paint called Dutch pink, is obtained from weld.

To save seed, select a few of the largest and healthiest plants, and leave them to ripen. The seed is easily separated. The chief disease of weld is the mildew, to which it is very liable when young, and this is the reason that it is often sown with other crops."

SUMACH (Rhus glabrum, R. coriaria and R. cotinus).

The Rhus glabrum is the common sumach of the United States which grows spontaneously on fertile soils. It is con siderably used by dyers, and the tanners of light leather. It is, however, much inferior to the *R. coriaria* or Sicilian sumach, which is imported into this country from Spain, Portugal, Sicily, Syria and elsewhere, and sells at from $50 to $120 per ton. It is a dwarf, bushy shrub, smaller than the American, but with much larger leaves. These with the seed cones and young stems, are all used by the manufacturers. The *R. cotinus* or Venice sumach, is the *fringe tree* or *burning bush,* a shrub for ornamental grounds, bearing a flossy, drab-colored blossom. It is known in England as *young fustic,* and is much used in the arts.

Cultivation and treatment.—All the sumachs are propagated by layers, though it is probable they might, under favorable circumstances, be raised from the seed. On good soils they grow in great profusion. The harvesting consists simply in cutting off the young branches with the leaves and seed cones attached, in clear weather, drying them thoroughly, without exposure to either rain or dew, and packing them in bales of about 160 lbs., for market.

The sumach is highly astringent, often taking the place of galls. This quality is much enhanced by warmth of climate ; and the most valuable article is brought from the more southern regions. There is no doubt this species of plants might be cultivated with great profit in the southern States, and thus save the large amount annually expended in its importation, which is constantly increasing. The total importation is now estimated at between one and two millions of dollars per annum.

THE TEASEL OR FULLER'S THISTLE (Dipsacus fullonum).

Is another article exclusively used by the manufacturers for the purpose of raising a nap, or combing out the fibres, upon the dressed surface of woollen cloth or flannels. The consumption cannot of course be extensive, being limited exclusively to this demand. There is but one kind cultivated. A bastard variety of spontaneous growth exists in portions of our middle States, which resembles the useful teasel, with this peculiar difference, that the ends of the awns or chaff on the heads are straight, instead of hooked, which renders them perfectly worthless.

Cultivation.—The teasel is a biennial, requiring two years to mature. It is sown on a deep, loamy clay, previously well plowed and harrowed, in drills 20 inches asunder, leaving a plant in every 10 inches ; or, if planted in hills, they may be about 16 inches apart. The ground should be kept light by occasional stirring, and free from weeds. The plants are generally stronger and more thrifty if allowed to mature where sown ; and to accomplish this, the intermediate spaces between the hills may be annually planted with new seed. Many adopt the plan of sowing in beds and transplanting.—Although hardy, there is sometimes an advantage in covering the young plants with straw during the winter, which can be conveniently done only when they are compactly placed in beds.

Gathering.—Those intended for use should be cut with a stem eight inches long below the head, just as it is going out of flower when the awns are the toughest; and as these mature at different times in the same plant, they should be cut successively as they come forward. Those intended for seed, which should always be the largest, strongest heads, must be suffered to remain till ripe, when they can be gathered and threshed with a flail. Spread the others thinly, and dry under cover where no moisture can reach them. They may then be assorted into three parcels, according to size and quality, and packed in large sacks, when they are ready for market. The crop on good soils well cultivated, may be stated at 150,000 to 200,000 per acre, worth from $1,50 to $2.00 per thousand.

The use of teasels has been to a considerable extent superseded in this country, within a few years, by the introduction of metallic nappers. These consist of thin, steel plates, with fine teeth arranged compactly on a shaft, forming a continuous cylinder of slightly projecting teeth, which are almost indestructible by use.

MUSTARD.

Fig. 52.   **Fig. 53.**

There are two species of mustard raised in the United States; *the white* (*Sinapis alba*, Fig. 52), which is most usually cultivated as a forage plant; and *the black* (*S. nigra*, Fig. 53), generally raised for the seed. It requires a rich, loamy soil, deeply plowed and well harrowed. It may be sown either broad-cast, in drills about two feet apart, or in hills. Mr. Parmelee, of Ohio, thus raised on 27 acres, 23,850 lbs., which brought in the Philadelphia market, $2,908; an average of over $100 per acre. The ground on which it is planted must be frequently stirred and kept clear of weeds. When matured, it should be carefully cut with the scythe or sickle, and if so ripe as to shed, laid into a wagon box with tight canvas over the bottom and sides, so as to prevent waste. As soon as it is perfectly dry, it may be threshed and cleaned, when it is ready for market.

The white mustard is a valuable crop as green food for cattle or sheep, or for plowing in as a fertilizer. For feeding, the white is much preferred to the black, as the seed of the latter is so tenacious of life, as to be eradicated with difficulty when once in the ground. The amount of seed required per acre is from eight to twenty quarts, according to the kind and quality of the land, and the mode of planting or sowing. It may be sown from early spring till August, for the northern and middle States, and till the latter part of September for the southern.

The crops yield from 25 to 30 bushels per acre. Both are excellent fertilizers for the soil.

THE HOP, (Humulus lupulus, Fig. 54).

There are several varieties of hops, indigenous to this

country. They grow best on a strong loam or well-drained clay, with light sub-soil. If the latter be retentive of water, the hop will soon dwindle or die. If made sufficiently rich, it will flourish on light loam or gravels; but a new, strong soil is better, and this requires little or no manure. The most desirable exposure is a gentle slope to the south: but this should be where there can be a free circulation of air amidst the tall vege-

Fig. 54.

table growth, which characterizes the luxuriant hop field beyond every other northern crop.

Cultivation.—If the land has been long in use, it should be dressed with compost and alkaline manures; or what is nearly equivalent, with fresh, barnyard manures, on a previously well-hoed crop, made perfectly free of all weeds, and deeply plowed and harrowed. Then mark out the ground at intervals of six feet each way and plant in the intersection of the furrows, and unless the ground be already rich enough, place three or four shovels of compost in each hill. The planting is done with the new roots taken from the old hills, which are laid bare by the plow. Each root should be six or eight inches long and must contain two or more eyes, one to form the root, and the other the vine. Six plants are put in a hill, all of which should be within the compass of about a foot, and covered to a depth of five inches, leaving the ground level when planted. The first season, the intermediate spaces between the hills may be planted with corn or potatoes, and the ground carefully cleared of weeds, and frequently stirred. No poles are necessary the first year, as the product will not repay the cost. The ground should receive a dressing of compost the following spring, and the plants be kept well hoed and clean.

Poles may be prepared at the rate of two or three to each hill, 20 to 24 feet long, and selected from a straight, smooth under-growth of tough and durable wood, from four to seven inches diameter at the butt end. These are sharpened and firmly set with an iron bar, or socket bar with a wooden handle, in such a position as will allow the fullest effect of the sun upon the hills or roots. When the plants have run to the length of three or four feet in the spring, train them around the poles, winding in the direction of the sun's course, and fasten below the second or third set of leaves, where there is sufficient strength of vine to sustain themselves. They may be confined with rushes, tough grass, or more easily with woollen yarn. This operation is needed again in a few days, to secure such as may have got loose by the winds or other causes, and to train up the new shoots.

The gathering of hops takes place when they have acquired a strong scent, at which time the seed becomes firm and brown, and the lowest leaves begin to change color. This precedes the frosts in September. The vines must first be cut at the surface of the ground, and the poles pulled up and laid in convenient piles, when they may be stripped of the hops, which are thrown into large, light baskets. Or the poles may be laid on long, slender boxes with handles at each end, (to admit of being carried by two persons), and as the hops are stripped they fall into the box. Be careful to select them free from leaves, stems and dirt.

Cultivating the second year.—After gathering in the fall, the hops should be hilled or covered with compost and all the vines removed. The following spring when the ground is dry, the surface is scraped from the hill, and additional compost is added, when a plow is run on four sides, as near as possible without injury to the plants. All the running roots are laid bare and cut with a sharp knife within two or three inches of the main root, and the latter are trimmed if spreading too far. It is well to break or twist down the first shoots and allow those which succeed to run, as they are likely to be stronger and more productive. Cutting should be avoided unless in a sunny day, as the profuse bleeding injures them. The poles will keep much longer, if laid away under cover till again wanted the following spring.

Curing or drying.—This is an important operation. It may be done by spreading the hops thinly in the shade and stirring them often enough to prevent heating. But when there is a large quantity, they can be safely cured only in a kiln. The following mode is recommended by Mr. Blanchard:

" For the convenience of putting the hops on the kiln, the side of a hill is generally chosen. The kiln should be dug out the same bigness at the bottom as at the top ; the side walls laid up perpendicularly, and filled in solid with stone, to give it a tunnel form. Twelve feet square at the top, two feet square at the bottom, and at least eight feet deep, is deemed a convenient size. Sills are laid on the top of the walls, having joists let into them like a floor ; on which laths, about one and a half inches wide are nailed, leaving open spaces between them three-fourths of an inch, over which a thin linen cloth is spread and nailed at the edges of the sills. A board about twelve inches wide is set up on each side of the kiln, on the inner edge of the sill, to form a bin to receive the hops. The larger the stones made use of in the construction of the kiln, the better; as it will give a more steady and dense heat. The inside of the kiln should be well plastered with mortar to make it completely air-tight. Charcoal made from yellow birch, sound hickory or maple is the only fuel proper to be used in drying hops. The kiln should be well heated before any hops are put on, and carefully attended, to keep a steady and regular heat. Fifty pounds of hops, when dried, is the largest quantity that should be dried at one time, on a kiln of this size, and unless necessary to put on that quantity, a less would dry better.

The green hops should be spread as evenly and as ligh‧ ᴛs possible over the kiln. The fire at first should be mode rate, but it may be increased as the hops dry and the steam is evaporated.

The hops, after lying a few days, will gather a partial moisture, called a sweat. The sweat will probably begin to subside in about eight days, at which time, and before the sweat is off, they ought to be bagged in clear dry weather. As the exact time when the hops will begin to sweat, and when the sweat will begin to subside or dry off (the proper time to bag them), will vary with the state of the atmosphere, it will be necessary to examine the hops from day to day, which is easily done by taking some of them from the centre of the heap. If on examination you find the hops to be very damp, and their color altering, which will be the case if they were not completely dried on the kiln, overhaul and dry them in the air.

Hops should not remain long in the bin or bag after they are picked, as they will very soon heat and become insipid. The hops should not be stirred on the kiln until they are completely and fully dried. Then remove them from the ᴋiln into a dry room and lay in a heap unstirred until bagged. This in done with a screw, having a box made of plank the size of the bag into which the cloth is laid and the hops screwed into the box, so constructed that the sides may be removed and the bag sewed together while in the press. The most convenient size for a bag is about five feet in length, and to contain about two hundred and fifty pounds. The best material is coarse, strong, domestic tow cloth ; next to that, Russia hemp bagging.

Those who have entered considerably into the cultivation of hops, build houses over their kilns, which are convenient in wet weather, otherwise, a kiln in the open air would be preferable. It is necessary to have these buildings well ventilated with doors and windows; and to have them kept open night and day, except in wet weather, and then shut those only which are necessary to keep out the rain. Or if a ventilator was put in the roof it would be found advantageous. I have seen many hops injured both in color and flavor by being dried in close buildings. Where the houses over the kilns are large for the purpose of storing the hops, make a close partition between the kilns and the room in which the hops are stored, to prevent the damp steam from the kilns as it will color them, and injure the flavor and quality."

Diseases.—Hops are liable to attack from various insects, blight, mildew, &c. There is no effective remedy of general application for either. The best preventives are new or fresh soil, which is rich in ashes and the inorganic manures, and in a fine tillable condition to insure a rapid growth, by which the hops may partially defy attack; and open planting, on such positions as will secure free circulation of air. When properly managed, hops are one of the most productive crops; but their very limited use will always make them a minor object of cultivation.

THE CASTOR BEAN, (Ricinus communis, usually called Palma Christi, Fig. 55).

Is a native of the West India islands, where it grows

FIG. 55.

with great luxuriance. It is cultivated as a field crop in our middle States, and in those bordering the Ohio River on the North. It likes a rich, mellow bed, and is planted and hoed like corn. It attains the height of five or six feet, and bears at the rate of 20 to 28 bushels per acre. The seed is separated from the pods, bruised and subjected to a great pressure, by which it yields nearly a gallon to the bushel, of *cold-pressed* castor oil, which is very much superior to that extracted by boiling and skimming. The last is done, either with or without, first slightly roasting. This oil forms not only a mild cathartic, but with some is an article of food. Its separation or conversion into a limpid oil for machinery and lamps, and into stearine for candles, has lately much increased its valuable uses.

ARROW ROOT (Maranta arundinacea, Fig. 56).

This plant is very extensively cultivated in South Ameri-

11

ca, the West India Islands and in Florida. It requires *a light, loamy, fertile soil* of good depth. *It is propagated* by dividing the roots and planting in drills, 12 to 18 inches apart. The ground requires stirring occasionally, and to be kept clear of weeds. When a year old, the roots are taken up, well washed, then reduced to a pulp by bruising or grinding. The pulp is then passed through a sieve, and after the fecula or starch has settled, the water is poured

Fig 5).

off. The sediment is again washed in pure water, then dried, in which condition it forms the arrow root of commerce. This constitutes a light digestible food for invalids, but affords little nourishment. It is essentially the same as potato flour and tapioca, or the product of Manchot or sweet cassava.

GINSING (Panax quinquefolium, Fig 57).

This plant is indigenous to the northern, middle and western States, where it grows spontaneously on the hill-sides when shaded by the forest trees. It yields numerous fleshy roots, of a yellowish color, from one to three inches long, which are dug, washed and dried, when they are ready for market. It has a sweetish and slightly aromatic taste, and contains considerable proportions of gum and starch. It possesses little merit as a medicine, though highly esteemed for its imaginary virtues by the Chinese. The shipments

Fig 57.

from this country for China have sometimes reached nearly half a million of dollars for a single year. It has not been cultivated to any extent in this country.

THE TEA PLANT (Thea bohea and T. viridis, Fig. 58).

FIG. 58.

This plant has been introduced to some extent, into various parts of the United States within the last few years. It grows extensively in China, between the latitudes 27° and 32°; and in the Island of Japan, it flourishes as far north as 45°.

It is propagated by planting two or three seeds together, at a distance of four or five feet apart each way, in the bed where they are to grow, in a dry, silicious soil, of moderate fertility, and generally on the hill-sides. In the northern provinces of China, the tea plant occupies a rich, sandy loam. It requires little attention, except to be kept clear of weeds. The leaves are plucked when the plant has attained a three years growth, and when dried, constitute the tea of commerce.

The leaves are picked three times in a season. The first, and but partially expanded leaflets, yield the best quality of tea, known in Europe and America as the *Imperial*. The next picking gives an inferior quality ; and the third yields the lowest in value. These are again subdivided, into an almost endless variety of sorts or *chops*.

The leaves are cured by heating them under cover, on iron pans, from which they are taken while hot and carefully rolled by hand. This operation is performed two or three times, and all the moisture thoroughly expelled, when it is assorted into various qualities and put up for sale or use.

The tea plant becomes unthrifty and stunted under the close harvesting of the leaves; and at the age of six to ten years, requires to be partially cut down, to secure a fresh growth of thrifty shoots. The immense and increasing consumption of this article will justify the fullest experiments at the South, with the view of adding this to our excessively limited list of southern staples. An extensive effort is at this moment made in Georgia and the Carolinas, by a gentleman from New York, for the establishment of tea planta tions in each of those States. I saw several thousand choice plants just imported from Europe, on their way to those States, and which, I learn, have been subsequently transplanted and are growing finely.

The plant has been cultivated in Brazil, France and Algiers, for many years, but it has succeeded only in the former country, to much extent. The soil of Algiers has been found too dry and the climate too hot ; while in France, little attention has thus far been devoted to it.

SILK.

This valuable product has been more or less an object of attention in this country, since its early settlement. It was raised on a limited scale in the then southern Provinces, long before their separation from Great Britain ; and for more than a century, good sewing silks have been made, to a small amount, in various places in New England. Occasionally, strong, domestic silk fabrics have been manufactured, which had the merits of comfort and almost perpetual durability, but with little pretension to style or conventional taste.

The enormous importations of silk into this country, in 1836, exceeding $20,000,000, awakened the attention of our countrymen to the great value of this material ; and a speculation in the morus multicaulis mulberry, at that time thought to be the best species for the silk worm, was the result, which, for a time, almost rivalled the tulip mania of Holland. The general effect, however, was beneficial. It scattered the material for the support of the silk worm throughout the country, and induced an attention to the rearing of this useful but humble servant of the pride and luxury of mankind, that might not have been realized to the same extent for many years subsequent. There is a large and increasing attention to this subject, but I regret to add, the production of our raw material is far below the demand in this country, while the manufactured article is largely imported.

My limits will not admit of minute directions for the management of a cocoonery, nor is this essential to a successful result. Moderate intelligence and skill, with close attention, will enable almost any one to produce the raw silk to a profit.

Varieties of the Mulberry for feeding.—The kinds of trees best suited to the health of the worm, and the weight and value of its product, are the *Alpine* and *Canton* The foliage of these is more solid and nutritious than the *multi-caulis*, and on rich or wet ground, they are far preferable to any other. The multicaulis, though very succulent and watery in moist land, and therefore liable to induce disease in the worms, is still a prolific and healthful food for them where grown on dry or upland soils. Many others of the mulberry family are more or less suited to the object ; and even some of the indigenous, uncultivated varieties of this country, have been found to answer a very good purpose for feeding, where other sources of supply had failed.

Manner of planting the Mulberry.—The usual system now adopted, is not to raise the trees in orchards to their full size as formerly practiced, but to sow or plant thickly, in drills or hills, and cut the sprouts and young branches as they shoot out. This gives an immediate return for the planting ; and it enables the person engaged in this enterprise, to commence his operations, without the long delay conse-quent upon the remote maturity of the trees.

Variety of worms.—The *peanut* is usually esteemed the best variety, being more hardy and productive, and yielding generally a finer quality of silk. There are several others, as the white, the sulphur, &c., which possess much merit.

Rearing and management.—The eggs must be kept in a cool, dry place, till ready to be hatched, the temperature not exceeding about 50° Far. When the mulberry leaves begin to make their appearance in the spring, the eggs may be exposed to a temperature of about 65°, then gradually raised to 75°. At this last degree of heat, from seven to ten days will be sufficient to hatch the eggs. .

Immediately commence feeding with fresh, but not wet leaves ; and supply them as wanted, till the worm has at-tained maturity and is ready to wind the cocoon. During the period of moulting, which occurs four times in the life of the worm, their customary food should be withheld from them. Some wilt the leaves partially before feeding, which is well enough as a preventive to an excess of water in their food.

The temperature of the room occupied by the worms should be kept at about 75°. A slight variation from this is not objectionable ; but where it varies materially, artifi-cial means must be resorted to for maintaining a nearly uni-form temperature. Cleanliness in feeding and removing the excrements and dead worms is important ; and there should be a free circulation of air, without exposing them to moisture and the depredations of birds, rats, or mice. Chloride of lime is an excellent purifier for the cocoonery.

When ready to spin, which may be known by their ceasing to eat, raising their heads, and clambering about their feeding boards, the worms may be supplied with poplar or other branches ; or wisps of straw tied at the upper end, and spread at the bottom, for the worm to crawl upon and attach his cocoon. The straw may be secured by bracing it between two shelves.

Breeding.—After remaining about eight days, a sufficient number of the best cocoons should be first selected for breeding. These must be nearly equal in the quantities of male and female worms, the first being generally indicated by a pointed end, and somewhat drawn in at the middle ; while the latter is nearly alike at either end. In about fourteen days the millers come forth and couple. After a connection of twelve hours, throw away the males, unless there is a deficiency, when they may be retained for further use ; but the progeny is not generally so strong as when the male is used but once. Soon after, the female voids a brownish matter, when she is placed on a clean paper or muslin, and put away in a dark place, where she lays her eggs and dies.

From 100 to 120 pairs of millers will produce an ounce of eggs. Each female lays from 300 to 500 eggs, averaging about 350. An ounce of eggs contains about 40,000. If well saved from good millers, and safely kept, they will nearly all hatch and produce good worms. Our climate is admirably adapted to the production of the silk-worm, as is shown by the fact, that while an average of 30 to 60 per cent. of the whole worms are lost in Europe, from climate, food, and irremediable disease, scarcely five per cent. are lost in this country, under careful management, from the same causes.

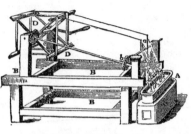

Piedmontese Reel. Fig. 59.

Reeling.—The cocoons may be reeled immediately after they are formed, if convenient. If to be kept for any time, the chrysalis must be stifled, which is done by exposure to a hot sun for two or three days, or baking in an oven at a temperature of about 200°.

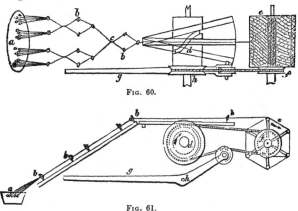

FIG. 60.

FIG. 61.

Fig. 59 is the most approved Piedmontese reel. Fig. 60 and 61 is a lately adopted French reel, represented in two views. A simple reel may, however, be made by any mechanic, that will answer the purpose for making domestic silk, but not if designed for sale. The reeling for market is a very nice operation, and requires a good reel and an experienced reeler, though both are easily procured with a little attention. When ready for reeling, place the cocoons in clear, soft water, raised nearly to the boiling point, then gently press them under with a light brush of broom-corn, and the fibres of the silk will adhere. After taking off the outer covering or tow, the silk is run rapidly on the reel, with enough threads to make the fibre of the required size. Keep the water pure by skimming, and changing as often as necessary. The silk is then allowed to dry immediately, when it may be packed for market.

CHAPTER XI.

FRUITS.

THE production of a variety of fruits, to the extent at least of his own wants, ought to occupy the attention of every farmer. The soil and climate of the United States, are almost everywhere suited to their cheap and easy propagation. They are a source of profit for the market, they are useful for stock, and they afford some of the choicest and most economical luxuries for domestic use. Success in their cultivation may at all times be secured, by a judicious selection of trees, soil and location, and by an intelligent and proper attention thereafter.

THE APPLE.

The locality for the apple orchard must depend entirely on the climate and soil. In warm latitudes, a northerly exposure is best when not subject to violent winds, as these from any quarter, are liable to blast the fruit while in blossom, and blow it from the tree before it is ripe. It is important to protect an orchard from the bleak winds which prevail in its immediate neighborhood, by a judicious selection of the ground. A warm and sunny exposure subjects the buds in spring to premature swelling, and these are often cut off by the severe spring frosts that follow; when a colder position would retard their budding until the season was sufficiently advanced for their protection.

Soil.—All the varieties of soil between a stiff, unyielding clay and a light, shifting sand, are friendly to the apple. The soil best suited to the perfection of fruit is a moist, friable, calcareous loam, slightly intermixed with fine gravel. This may run either into a sandy loam, which usually rests upon a sub-soil of sand or gravel, or into a clayey loam with a sub-soil of stiff clay. Either of these is a good soil for the orchard. The ground should be rich enough for the production of good crops of grain, roots or grass. This degree

11*

of fertility is absolutely necessary for the thrifty growth of the tree, and its existence in a healthy and vigorous state. Springy or wet land is decidedly objectionable, and if the farmer can appropriate no other for this purpose, it should be well drained, either by under-ground ditches or open trenches, sufficiently deep to carry off the water for a depth at least of two feet below the surface, so as to leave the soil which is perforated by the roots, in a warm and active state. Rocky and stony soils of the above descriptions, are usually well suited to the growth of fruit trees. The stones keep the ground moist, loose and light. Some of the finest fruits grow where there is scarcely room to place the roots of the tree between the rocks. But a sufficient area of earth is necessary for an ample growth of wood, and the full size of the tree at maturity.

Stiff clays and light blowing sands, under very nice cultivation, will grow fruits; but they require active manures. Clays should be often plowed, particularly in the fall, that the soil may be ameliorated by the winter frosts. The sands require compact culture and appropriate manures. All such as are suited to ordinary crops on these lands, will promote the growth of trees. But it is preferable to appropriate soils more suitble for the orchard, as the fruit will be larger, fairer and better flavored, and the trees of much longer duration.

Planting.—The soil should be prepared by deep plowing, before planting the trees. The sub-soil plow will accomplish this more effectually than can otherwise be done. Then dig the holes from three to six feet in diameter and twelve to eighteen inches deep, according to the kind of soil and the size of the tree. The more compact the soil the deeper and larger should be the hole. When ready to plant, let enough of the best or top soil be thrown into the bottom of the hole, so that the tree may stand about one inch lower than when removed from the nursery. Take up the tree so as to injure the roots as little as possible. If any be broken cut them off, either square or obliquely, with a fine saw or sharp knife. When left in a bruised or broken condition they will canker and decay in the ground; but if thus cut off, numerous rootlets will spring out at the termination of the amputated root, which strike into the soft earth and give increased support to the tree. Should the soil be poor, the roots must be covered and the holes filled with good earth. If the hole be small, the surrounding land hard, and the roots bent up and cramped, the tree cannot

11*

grow ; but if it finally survives after a long time of doubt and delay, it creeps along with a snail's pace, making little return to the owner. When the tree is crooked, confine it with a straw band to a stake firmly planted in the ground. This is the best ligature, as it does not cut the bark, which small cords often do, and it gradually gives way as the tree increases in size. When thus planted, well manured and well looked after, the tree thrives, and in a few years, rewards the owner with its delicious and abundant fruit.

The season of planting may be any time after the fall of the leaf in autumn, till its re-appearance in the spring, provided the ground be not frozen. Early spring is to be preferred for planting stone fruits. They may be removed while in embryo leaf and blossom with entire success, but it is better to do this before the bud is much swollen. If one time be equally convenient with another, fall planting is to be preferred for fruit generally, as the earth then becomes settled about the roots early in the following season. This is particularly advantageous when the spring is succeeded by a severe summer's drought. The transplanting of trees is an operation of the greatest importance to their success More fruit may be reasonably anticipated for the first ten years, if not forever, from one tree well planted, than from three indifferently done.

It sometimes occurs in removing trees from a distance, that they arrive at their destination after the ground is frozen. In such cases, a trench should be dug in soft earth and the trees laid in it, at an angle of about 45°, three or four inches apart, the roots carefully placed to prevent breaking, and the earth piled on them for a foot up the trunk, and eight or ten inches over the roots. This will preserve them until spring without detriment to their future growth, and it is often done by nurserymen and others, who remove their trees from one location to another without loss. Apple trees should never be planted in the orchard at a less distance than from thirty to forty feet ; the distance to depend on the fertility of the soil and the kind of tree, some growing much larger and throwing out their branches more laterally than others. If too near, the trees do not receive the requisite quantity of sun and a free circulation of air, both of which are essential to the size, flavor and perfection of fruit.

Cultivation.—A previously uncultivated or virgin soil is the best for an orchard ; but if such is not available, then such as has been long in pasture or meadow is most suitable. The most efficient manures are swamp muck, decayed leaves and vegetables, rotten wood, chip manure, lime, ashes, gypsum and charcoal. Trees draw their food mostly from the soil, and to supply the elements of their growth in abundance, the earth must occasionally be renewed with those materials which may have become partially or wholly exhausted. When carefully plowed and cultivated in hoed crops, orchards thrive most rapidly, if care be taken to protect the trees from damage either to the trunks or roots. All tearing of the roots is objectionable. The ground should be kept rich and open, so as to be pervious to the rains, the sun, and the atmosphere. Under these conditions the trees will thrive vigorously.

When lands are in meadow, a space of three to six feet in diameter around the trunk, according to the age and size of the tree, ought always to be kept free from turf. Pastures are so bared by the tread of animals, and the closeness of their cropping, that the roots of the trees get their share of benefit from the sun and rains. From this cause, pastures are better suited to orchards than mowing lands; for the latter are so completely covered by the rank growth of grass, that the tree suffers, and without the aid of manures, and the annual loosening of the ground for a few feet around, the tree in some cases dies from exhaustion. All kinds of cereal grains are bad for orchards, except, perhaps, buck-wheat. The preparation of the ground for this crop, by early summer plowing, is highly conducive to the growth of trees; and its nutriment being drawn largely from the air, it robs the roots of a small amount only of the materials in the soil.

A neighboring farmer, whose management many years since came under my notice, had a small mowing lot adjoining his barn and cattle sheds, which was surrounded with a stone wall. The soil was a moist, gravelly loam, every way fitted for the growth of the apple, as was shown by there having been several flourishing orchards on similar soils in the immediate vicinity. He filled this with apple trees set in small holes at the proper distances, the rows terminating close to the wall on each side, and also near his barn and sheds. After setting out, the trees were staked, and then left to grow as best they could without further cultivation. Those remote from the wall and buildings remained stationary for several years, while those under their influence, after two or three years, began to show a vigorous growth. The grass was ren oved annually, and the trees received no cultivation, save perhaps a bushel or two of chip manure occasionally thrown around them. Twenty years after they were planted, the trees next to the wall and buildings were thrifty and had attained a large size, while many of the others had died, a few had grown to one fourth the size of the outer ones, and others were still smaller, mossy, and showing signs of a premature old age. Not one third of the trees gave any return of fruit. The wall and buildings kept the soil next them light and moist, while that in the more open field spent all its energy upon the grass. To make an orchard profitable, the soil must be properly cultivated, till the trees have attained a considerable growth, and show so much vigor and thrift, that their expanded roots may be safely left to provide their own nutriment.

Pruning.—This operation must commence at the planting of the tree, the top of which should always be in proportion to the size and number of the roots. If the top be high and spindling, shorten it so as to throw the lateral shoots into a graceful and branching form The limbs may commence about six feet from the ground. Pruning should be done annually, as the labor is then trifling; and the expenditure of vital force in maturing wood which is afterwards to be cut off, is thus saved, and the branches to be removed being small, the wounds readily heal. In this case, no covering is required for the wound, as one season's growth will heal it. The top should be sufficiently open to admit the sun and air.

The best time for trimming is when the tree is in bloom, and the sap in full flow. The proper instrument is a fine saw or sharp knife, and the limb should be cut off close to the remaining branch. The sap at this time is active, and is readily converted into new bark and wood, which speedily forms over the cut. But this is a busy season with the farmer, and if he cannot then prune his trees, he may do it when more convenient, taking care to secure the wounds by an efficient covering of salve. Old trees, or such as are growing vigorously and have been long neglected, often require severe trimming, which should always be done in May or June; and when the wounds are large, they must be covered with a coat of thick, Spanish-brown paint or grafting wax. If they are left exposed, and the growth of the tree be slow, decay will often take place before they are healed Too much care cannot be used in these operations. In large trees, a ladder ought always to be at hand, to avoid breaking the limbs by the weight of the operator. If by too close planting the branches of different trees be brought into contact, thorough pruning is absolutely necessary, as without it, good fruit cannot be obtained.

Grafting and budding.—These operations are so simple, and usually so well known by some individual in every farming neighborhood, that no description of either operation is necessary. *Grafting wax* of the best kind is made with four parts of rosin, one of tallow and one of beeswax, melted and stirred together, then poured into a vessel of cold water. As soon as cool enough, work and draw it out by hand, like shoe-makers' wax, until it is entirely pliable. It may then be used immediately, or laid up and kept for years. The mode of applying it is known to every grafter.

Scions must be the growth of the preceding year, and cut from well ripened, thrifty wood, in the months of January, February or March, before the buds begin to swell with the flow of the spring sap. Tie them up and keep in a moist, cool place, a cellar bottom, or box of moss or earth till ready for use. When circumstances require it, grafts may be cut at any time after the fall of the leaf, but the months indicated are best in all localities north of lat. 40°. The best time for budding is in July and August. This should be done while the sap is in flow and the bark is loose, as at no other time is success certain.

Selection of trees.—Select these from seedlings. Suckers from the roots of mature trees are objectionable, as tending to throw up suckers themselves, which are always troublesome. When they appear, cut them close to the root or stem, and if properly done they will rarely sprout anew.

Planting the seed.—If the farmer wish to raise his own trees, he can sow the seed or pomace in rows in the fall. After they come up in the spring, weed and hoe them like any vegetable. When a year old they should be carefully taken up, the tap root cut off and replanted in rows four feet apart, and at least a foot distant in the rows. They should be regularly cultivated till they are one and a half or two inches diameter at the base, at which time they are fit for the orchard. These operations are, however, the appropriate business of the nurseryman, for whose guidance there should always be at hand some standard work on the cultivation of fruits. Of these, Kenrick's, Downing's and Hovey's are at present the best American treatises.

Gathering and preserving.—For immediate use, apples may be shaken from the tree. For winter consumption or packing for market, they should be carefully picked by hand with the aid of ladders, to avoid bruising the fruit and injuring the limbs. To preserve apples, the best method is to lay them carefully into tight barrels or boxes immediately after picking, with a thin layer of perfectly dry chaff on the bottom, and after being lightly shaken together, another layer of chaff on the top may be added, though this is not essential. They may then be tightly headed or covered so as to exclude the air. Then put the boxes or barrels away into a dry place, and keep as cold as possible above the freezing point. But if slightly frozen, they will not be injured if suffered to remain unpacked till the frost leaves them. When thus managed, they will keep as long as they are capable of preservation. Bins in the cellar are good for ordinary use, if closely covered. When exposed to the air, warmth or moisture, apples soon decay. If too dry, they wilt and become tasteless. They are sometimes buried in the earth like potatoes, but this impairs the flavor and gives them an earthy taste ; and they seldom keep so well after removal in the spring as when they have been stored in barrels.

For farm stock, apples are healthful and fattening, and the better the quality of fruit the more valuable are they for this object. A variety of both sweet and sub-acid should be cultivated. The saccharine matter of the apple is essentially the fattening property, and this abounds in some kinds of the sub-acid. Animals like a change in their food as well as man, and both sweet and sour may be fed to them alternately. When the soil and climate are adapted to them, food from apples can probably be more cheaply supplied to stock in the northern States, than from any other plants of artificial cultivation, excepting grass and clover. Swine have been often fatted upon them with an occasional change to grain ; and when fed to horses, cattle and sheep, with hay, they are almost equivalent to roots. That tree must be badly cultivated, which in ten years after planting, will not produce five bushels of apples in a season, and these at ten cents a bushel give an annual revenue of fifty cents a tree, or twenty dollars per acre for stock-feeding alone. At twenty years old, the tree will double that product casualties excepted; and as this estimate is based on their least valuable use, an increased profit may be anticipated from their conversion to other purposes. Good apples are rarely worth less than twenty-five cents a bushel; often three o four times that amount. The presence of swine among any kind of fruit trees, greatly conduces to their thrift. Besides the support of the swine derived from the fruit, their consumption of windfalls secures the destruction of such insects as are injurious to the trees or fruit, and the manure they drop, together with the loosening of the earth, resulting from constant rooting and the tread of their sharp hoofs, is of essential advantage to the growth and healthfulness of the trees. Sheep, turkeys, ducks and chickens answer the same purpose in a considerable degree, when suffered to frequent the orchards in sufficient numbers.

Making cider.—Good fruit is indispensable to the making of good cider. The suitable time for grinding is in October and November, and apples designed to be thus appropriated should ripen in these months. Such as are slightly acid are preferable for this purpose. As far as practicable, the fruit should be of one kind, fully ripe, but sound and undecayed. The mill must be thoroughly cleansed with hot water, and capable of grinding the pomace fine. This should lie in the vat at least forty-eight hours after grinding, and be turned once or twice before its removal into the cheese. Pomace so exposed, absorbs large quantities of oxygen, thus undergoing a necessary preparation for its conversion into good cider. All fruits are subject to this change, to a certain extent, just before ripening. When their juices are expressed or the pulp broken and exposed to the air, this effect is increased, and constitutes the *saccharine fermentation*. In both cases, the result is to increase the palatable and nutritive properties of the fruit, by converting their starch, gum and other vegetable matters into sugar.

When the pomace has been sufficiently pressed, it may be fed to cattle, sheep or swine, and the liquor put into barrels under cover, and allowed to remain till the pulp or feculant matter has been thrown out at the bung; and to aid its removal the barrel should be kept full. The second fermentation is the *vinous*, and by it a portion of alcohol is developed. This fermentation is slowly continued afterwards in the enclosed cask, until it reaches from six to nine per cent. When fermentation apparently subsides, take a clean cask, in which a small quantity of sulphur has been burned, to arrest any subsequent tendency of the liquid to change, draw the cider into this and bung tightly to exclude the air. The addition of charcoal, raisins, mustard seed or fresh meat produces the same effect as the ignited sulphur. After standing two or three months, closely confined, and in a cool place, it may be drawn off and tightly bottled for use. Its long preservation and improvement will depend on its being kept cool and well corked. In addition to its possessing a small proportion of alcohol, it then contains large quantities of carbonic acid gas, which occasions its rapid effervescence when uncorked, and gives to it that peculiarly pungent and agreeable flavor, so highly relished in the best specimens of the Newark cider.

Vinegar.—If the cider be allowed to remain in the cask in which it is first placed, and exposed to a warm temperature, it continues greedily to absorb oxygen, and quickly undergoes another fermentation, called the *acetic*, by which it is converted into vinegar. If intended solely for this purpose, the best and richest fruits give the strongest, best-flavored and soundest (most reliable) vinegar. When it has acquired its perfection, the vinegar should be kept air-tight and at a low temperature.

Best varieties of apples for cultivation.—Almost every section of the apple-growing regions of America, has a greater or less variety peculiar to itself ; and their valuable properties appear more fully developed in these localities than when removed to others. Such should of course be retained when of extraordinary excellence. There are some, however, which are of more general cultivation, cosmopolites throughout the apple climates, of fine quality. and possessing all the excellence of which the genus is capable. Thirty different kinds for each section or State will probably include all which it is desirable to cultivate, and for any one location perhaps twenty is sufficient. I mention below, the names of 30 standard varieties, all of which are now in successful cultivation in different parts of the United States and Canada. They are described by Downing, in his late work on the fruit trees of America, 1845.

Summer Apples.—Early Harvest, Red Astracan, Large Yellow Bough, Williams' Favorite.

Autumn Apples.—Golden Sweet, Fall Pippin, Gravenstein, Jersey Sweeting, Pumpkin Russet (by some, the Bellebonne), and Rambo.

Winter Apples.--Westfield Seek-no-farther, Baldwin,

Black Apple, Yellow Belle fleur, Detroit, Hubbardston Nonesuch, Green and Yellow Newtown Pippin, Northern Spy, Blue Pearmain, Peck's Pleasant, Rhode Island Green ing, American Golden Russet, English Russet, Roxbury Russet, Swaar, Ladies' Sweeting, Talman's Sweeting, Eso pus Spitzenberg, Waxen Apple, Wine Apple.

THE PEAR.

The pear is one of the most luscious, wholesome and pro-fitable of the market fruits, though not comparable to the apple for variety and general use. In a good soil and under proper cultivation, it is both vigorous and hardy. It is bud-ded and grafted like the apple, and requires the same treat-ment; it is as easy of propagation, frequently attains a greater size and age, and although longer arriving at matu-rity, it is a more abundant bearer. Its favorite soil is a clay loam. It needs little pruning, but usually it throws out an upright, graceful head, free from excessive bushiness. The trees may be planted 30 feet apart, an abundance of sun being requisite to full bearing and the perfection of the fruit.

Diseases.—The pear is seldom subject to more than one formidable disease, the fire blight, and to this, it is more ex-posed in some localities than others. The disease manifests itself generally in mid-summer, in the sudden withering of the leaves on one or more branches. The only effectual remedy is to cut off and burn the diseased limb, immediately upon its discovery. The causes are imperfectly known, but it has been variously ascribed to the presence of minute in-sects, to the excessive flow of sap, and to the severity of the winter.

Gathering and preserving the fruit.—Many pears re-quire to be picked just before they are ripe, and allowed to mature in the shade. They thus acquire a rich, juicy character they would not otherwise attain. Those intended for market or for long keeping, should be hand-picked and laid in a cool place ; and when perfectly dry, put up in casks like apples. Winter pears may be packed for preservation like winter apples.

The varieties to be selected depend entirely on the object of their cultivation. For market, the best and most popu-lar kinds only should be chosen ; and for family use, an equally good selection should be made of those maturing throughout the entire season.

I subjoin in their order of ripening, a dozen choice kinds,

the cultivation of which has thus far been thoroughly successful, and the qualities universally approved. The most of these are pears of American origin, which are to be preferred as promising more durability, hardiness and perfect adaptation to our climate and soils. I quote from Downing on fruits.

Summer and Early Autumn Pears.—Bloodgood, Dearborn's Seedlings, Bartlett or Williams' Bon Chretien, Stevens' Genesee.

Autumn Pears.—Beurre Diel, Seckel, Dix, White Doyenne or Virgalieu, Duchess D'Angouleme.

Winter Pears.—Beurre D'Aramberg, Columbia, Winter Nelis, Prince's St. Germain.

THE QUINCE.

This is also a valuable market fruit. It makes a rich, highly-flavored sweetmeat, and to this use it is entirely limited. The tree is easily raised by suckers and cuttings, and should be planted fifteen feet apart, in a rich, warm, heavy soil (a clayey loam is the best), rather moist, and in a sunny exposure where it will be sheltered from cold and severe winds. The wash of a barn-yard is its best manure, and it repays equally with the apple, for good cultivation. The fruit is large, sometimes weighing a pound, of a rich, golden color, and generally free from worms and other imperfections. It ripens in October and November. The orange quince is the best variety for common cultivation. The tree requires but little pruning. The trunk may be entire for two or three feet, or branch from the ground by two or more stems. The top should be kept open to admit the sun and air, and the trunk freed from suckers. When thus treated, it will live long and produce abundantly.

THE CHERRY.

Aside from the value of its fruit, the cherry is an ornamental shade tree, hardy and vigorous in its growth, and easy of propagation. It should be planted like the apple. For culinary purposes, the common red cherry is perhaps the best. This may stand sixteen or twenty feet apart, according to the soil and situation. The large Mazard or the English cherry requires more room, and if on a deep, warm, sandy loam, its favorite soil, it should be planted two rods apart as it grows to a large size. It will flourish luxuriantly on a clay loam or an open gravel, provided the soil be

rich and deep; but on these, it demands more careful cultivation. It seldom requires much pruning. Care must be
used with this as with all other fruit trees, to give it an open
head and to keep the limbs from crossing and chafing each
other. The varieties most in use are the *Common Red,
Kentish or Pie Cherry*, almost universally cultivated, the
English Mayduke, Black Tartarian (Graffion or Yellow
Spanish,) the *large Red Bigarreau, Elton, Belle de Choisy*
and the *late Duke.* These will form a succession of six
weeks in ripening and embrace their entire season. The
cherry is remarkably free from disease, and usually requires
but ordinary care in its cultivation.

THE PLUM.

The plum affords some of the most delicious of our cultivated fruits. It prefers a strong clay loam, but does well in
nearly all soils, except a light sand. It should be planted
like the apple, though on a more limited scale, as it has a
smaller and less vigorous growth. The proper distance is
sixteen to twenty feet. There are two formidable impedi-
ments in the cultivation of the plum. One is an insect,
which attacks the wood, and deposits its egg in the smaller
branches. This is followed by a large swelling or excres-
cence and if suffered to remain, will soon destroy its productiveness. The surest remedy is to cut off the branch at
once and burn it.

The Curculio commits its depredations on the young
fruit, soon after the blossoms disappear. These are frequently so destructive as to kill the fruit of an entire orchard.
Several methods of destroying them have been suggested,
of which the most simple and effectual is, to plant the trees
in such places as will admit the swine and poultry to feed
upon the fallen fruit and insects. Salt sprinkled around the
tree in the spring, is said to destroy them. The smoke of
rotten wood, leaves and rubbish which have been burned
under the trees when in blossom, has sometimes proved
beneficial. Paving the earth under the limbs to prevent the
burrowing of the insects, and some other remedies are recommended. This is a serious evil, requiring more observation and experiment than it has yet received.

Varieties.—The common Blue or Horse plum is cultiva
ted in numerous sub-varieties. Some of these are very good
others utterly worthless. Good plums are as easily raised
as poor ones, and these only ought to be cultivated. Young

trees bearing an indifferent fruit, can be headed down and grafted as readily as apples, but this requires to be done a month earlier in the spring, and before the buds begin to swell. The best kinds are the Yellow, Green, Autumn, Bleecker's, Imperial, Prince's Yellow, Frost, Purple and the Red Gages; Coe's Golden Drop, the Jefferson, the Grange, the Washington, the Columbia, Smith's Orleans, and the Red Magnum Bonum.

This last variety is more liable to the attacks of the curculio than many others. But its vigorous growth, great productiveness when not attacked, and its excellent quality for the table render it a desirable fruit. *For drying*, the German prune is perhaps the best, although several of the plums above named answer an excellent purpose. I have enumerated a larger variety of plums from the difficulty of cultivating the peach successfully in many parts of the northern States. They ripen nearly at the same time, and though not as delicious nor generally as popular, they are the best substitutes for it. Although liable to several diseases, the plum is more hardy and durable than the peach, and its cultivation is comparatively easy.

THE PEACH.

In the early settlement of our country and on virgin soil, the peach was easily propagated, free from disease, an abundant bearer and comparatively long-lived. If we except the first feature in its early history, we shall find it generally, differing widely in each of the others at the present day. It has become subject to so many casualties, as to have been almost entirely discarded in large sections of the United States, where it once flourished in the highest perfection. It is now most frequently reared on an extensive scale for market, by those who make it an exclusive business.

Its favorite soil is a light, warm, sandy or gravelly loam, in a sunny exposure, protected from severe bleak winds. Thus situated and in favorable latitudes, it grows with great luxuriance and produces the most luscious fruit. In western New York, and on most of the southern borders of the great lakes, the peach grows more vigorously and lives longer than in any other sections of the United States, frequently lasting 20 to 30 years, and bearing constantly and in abundance. Peaches are produced in immense quantities on the light soils near the Atlantic coast, in the States of New Jersey and Delaware. The crop of a single proprietor often amounts to $5,000, and sometimes exceeds $20,000 annually.

None but the choicest kinds are there cultivated, and these are innoculated upon the seedling when a year old. They are transplanted at two and three, and are worn out, cut down and burned at the age of from six to twelve years. The proper distance for them to stand, is sixteen to twenty feet apart, according to situation, soil and exposure. Constant cultivation of the ground is necessary for their best growth and bearing.

Diseases.—The peach is liable to many diseases, and to the depredations of numerous enemies. *The Yellows* is the most fatal in its attacks, and this can only be checked by the immediate removal of the diseased tree from the orchard *Of the insects*, the grub or peach worm is the most destructive. It punctures the bark, and lays its egg beneath it at the surface of the earth. When discovered, it should be killed with a penknife or pointed wire. A good preventive is to form a cone of earth a foot high around the trunk about the first of June; or if made of leached ashes, it would be better. Remove this heap in October, and the bark wil. harden below the reach of the fly the following year.

Varieties.—The best kinds in succession, from early to late, are the Red and Yellow Rareripes, Malacatune, Early York, Early Tillotson, George the Fourth, Morris' Red and White Rareripes, Malta and Royal George. These succeed each other from August to October.

THE APRICOT AND NECTARINE.

These are of the peach family, but generally inferior as a fruit and much more difficult of cultivation, being peculiarly liable to casualties and insects. They require the same kinds of soil and cultivation as the peach, with a warmer exposure. As they are propagated solely as an article of luxury and are not wanted for general use, I omit further notice of them.

THE OLIVE (Olea europæa).

This, next to the fig and gopher, is the earliest tree mentioned in history. (Gen. 8, 11.) It was ever a favorite with sacred and profane writers, and it is consecrated by both, and among all nations wherever cultivated, as the cherished emblem of peace, prosperity and abundance.

It grows spontaneously in the temperate regions of Asia

and Africa ; and from the earliest period, it has been propagated in the southern part of Europe. It has contributed largely to the support of the human race, in every age of the world; and it continues at the present time, to yield large quantities of food, materials for the arts, and immense revenues wherever it is made an object of attention. The small kingdom of Naples exports yearly 7,000,000 gallons of olive oil. The value of the annual production in France as early as 1788, was $15,000,000; yet she has since increased her consumption by importation from abroad, to the extent of $6,000,000 in a single year.

The olive was introduced into the Carolinas soon after their first settlement by European emigrants. Richard Blome, an old writer who describes the country in 1678, says, "the olive trees brought from Portugal and Bermuda increase exceedingly, and will produce a quantity of oil." Gov. Glum, states that "he lost an olive by intense frost in January, 1747, of 18 inches diameter;" and Dr. Milligan, of Charleston, adds in 1763, "we have plenty of olives."

The tree is easily propagated by cuttings, layers or seeds; and it is a hardy, self-sustaining tree when not exposed to very severe frosts. It will even remain uninjured in this country, under an exposure that cuts down the sweet orange and other trees reared among us successfully. It may perhaps, find obstacles to its productiveness in our protracted, sultry weather, remote from the sea; yet we have a long belt of coast and sterile hilly lands at the South, adapted to its growth, over much of whose area it will undoubtedly flourish. It loves a thin, dry, calcareous soil, and when once planted, it will, with little care, continue to yield large annual returns for centuries. The olive is among the longest-lived trees, whether wild or domesticated, and though hitherto comparatively but little cultivated among us, is worthy of the particular attention of southern agriculturists.

Uses.—The olive furnishes more fixed oil than all other vegetables combined, and this is of very extensive use, both in the arts and as an article of diet. The fruit is crushed in a mill and reduced to a pulp, and is then subjected to pressure. The oil floats on the surface of the expressed liquid, and after being drawn off and clarified, is fit for use. A portion of the best is bottled for the table, and the remainder is put up in barrels for coarser uses. A few comparatively, are reserved for pickling, and are much relished by those accustomed to them.

THE ORANGE.

The wild or bitter orange is found in Florida, in groves sometimes of thirty miles in extent. From its existing in no other part of North America, in its natural state, it is supposed to have been planted there by the Spaniards, in the early settlement of the country. This is the more probable, from its ready propagation, and the deterioration of the fruit from neglect. These native stocks are hardy, and are usually taken for grafting, or more frequently, innoculating with the best cultivated varieties. They afford a taller, hardier tree, and more vigorous and glossy foliage than the cultivated. They bear a profusion of fruit which, however, is worthless.

There are many choice kinds, among which, are *the Mandarin, the Navel, the Chinese, the Blood-red, the Sweet-skinned, the Pear-shaped and the Seedless.* They grow in profusion in Louisiana, Florida, Georgia, S. Carolina and Texas, where they afford a remunerating crop to the planter. *The soil* most propitious to the orange, is moist yet not wet, varying between a loamy sand and light clay, but loose and fertile.

They are propagated by layers, suckers or seeds; but are usually innoculated or grafted. They require little attention except to guard them from frost where the climate is severe. Most of the orange groves in Florida, were killed down to the roots by severe frosts, in the winter of 1834 and '35. But they generally withstand the effects of frost near the Gulf coast. The worst enemy to the orange tree in this country, has been the *Coccidæ* or *bark lice,* which fasten themselves by a thin pellicle or covering to the bark, beneath which a family of these insects grow up, and gradually extend their progeny by the same means, over the whole trunk and branches No effectual remedy has been discovered against their ravages; and many extensive plantations have been cut to the ground, in the hope that fresh shoots and a more vigorous growth would withstand their deleterious effects.

The produce from the orange tree is large, averaging in the best specimens, under favorable circumstances, 2,000 per annum. "Mr. Alvarez had a tree on the St. Johns, in Florida, which, in 1829, produced 6,500; and one tree in the same neighborhood, is said to have yielded 10,000 in one year." (*Browne's Trees of America.*) They yield a luscious healthful fruit, and their cultivation should be extended wherever the soil and climate will admit of their successful propagation.

THE FIG.

The fig is among the earliest cultivated and most popular fruits. It existed in Paradise, and its fruit was undoubtedly among the choicest that ministered to the support of our first parents. We read of it in the oldest historical writers, both sacred and profane; and its popularity has descended to the present day, wherever the soil and climate admit of its cultivation.

It is readily propagated by suckers and cuttings, from either roots or branches, or it may be grown from the seed. It grows rapidly and with slight attention, but seldom reaches a large size. From 20 to 25 feet in height and diameter, may be deemed the average size of American trees; but they have been known elsewhere, when arrived at an advanced age, to have doubled these dimensions.

The Banyan Tree of India (Ficus benghalensis), that propagates itself by dropping its extended branches to the ground, where they take root, and themselves become new trunks, for their further almost indefinite extension, till they overspread acres, is a species of fig, bearing an edible fruit of the size of a hickory nut. This and some other of the species furnish a gum from their sap, somewhat resembling caoutchouc or India rubber.

The soil best suited to the fig, is light, rich and moist. It is subject to few diseases and casualties, except from the scorching effects of too hot a sun. It loves a moist climate, and flourishes best in the neighborhood of the sea coast.

The fig is extensively cultivated in the Gulf States, and grows luxuriantly in the open air, as far north as 36°; and I have seen some of the hardier kinds flourishing in the open grounds near Baltimore. When protected by walls in summer and the trunks guarded properly in winter, it matures its fruit in New York. Florida, southern Mississippi, Louisiana and Texas, may be considered its most productive locality within the Union. Here it flourishes with scarcely any attention, and is subject to few casualties or diseases The fruit ripens from June to October, according to the soil situation, and the varieties reared. They usually produce two crops in one season, and in its most southerly habitat, three. The fruit pushes directly out from the branches and twigs, apparently without blossoming, though there is a minute flower concealed within the fruit.

The varieties are almost innumerable, and most of them delicious when grown under favorable circumstances. They are also healthful, and unlike the orange, are a substantial food. In other countries, among such people as subsist on light, meagre diet, they materially contribute to the sustenance of the inhabitants. They are a gentle laxative, and are frequently used as a mild emollient.

Their production in this country has been limited almost entirely to their use while fresh. We lack the long-continued, hot, dry weather necessary to cure them in the open air. It is not improbable, that artificial means may be used for preparing them for profitable export to those sections of the country where they are not raised. In this view, they may be deemed an important addition to the exports of our most southern territory, at some future day.

THE GRAPE.

The details for the proper rearing of this fruit demand a volume, but I can only refer to some prominent points in its cultivation. It grows wild and in abundance in many parts of the United States, and of tolerable quality, climbing over trees, rocks and fences in great luxuriance. I have seen in the eastern States, a dozen excellent native varieties of white, black and purple, of different sizes, shapes and flavor, growing within the space of a single furlong. So abundant were the clustering vines on the Atlantic coast in the vicinity of Narraganset Bay, that the old Northmen who discovered, and for a short time occupied the country in the 12th century, gave it the appropriate name of *Vinland* or *the Land of Vines.* The choicer kinds require loose, marly soils, with warm, sunny exposures and proper trimming. Thus cultivated, they are often raised with profit. The more choice and delicate kinds of the imported varieties, must have protection in winter and glass heat in summer, and are therefore only suited to a well-arranged conservatory.

Varieties.—The best American kinds are the Isabella and Catawba for the middle, and the Scuppernong for the southern States. North of latitude 41° 30, neither of the two former ripen certainly, except in long and warm seasons; and it is better for the cultivator within this range, to select some of the hardiest and best wild grapes of his own latitude, for out door propagation. Grafting a foreign variety

12

on a hardy, native stock, has been found to give a choice fruit, in great abundance, and with more certainty than could be secured by an entire exotic. Of the European, the varieties of Chasselas, Black Hamburgh, and White Muscat of Alexandria, are the best. In a good grapery either with or without artificial heat and proper attention, these can undoubtedly be raised at a price, which would yield to the horticulturist an adequate return ; and for this purpose, they are the best kinds to propagate, furnishing a long succession of fruit in its finest variety.

THE CURRANT

Is the first in importance of the small garden fruits. In the culinary department it has many valuable uses ; and it is a wholesome and delicious fruit when ripe. It grows with the greatest certainty and luxuriance, either from the suckers or cuttings. · The ground should be rich and well worked, and the bushes set at least six feet apart. They require plenty of sun and air like all other fruits. The red is the most common kind, but the large Dutch White is sweeter and more delicious, a great bearer, larger and as easily cultivated. The English Black is very productive, of great size, and makes a fine jelly. It has peculiar efficacy in sickness. The usual mode of planting currants near fences, is objectionable. They should stand out where the gardener can get around them, and be properly trimmed, or they soon get too thick. This improves the fruit, and insects and vermin are more effectually prevented from harboring among the bushes.

THE GOOSEBERRY.

This makes a palatable tart, and as a ripe fruit, possesses some excellence. It is easily raised, and prefers a cool, moist and rich soil in a sheltered position. It has been brought to the highest perfection in the north of England and Scotland, under the influence of their cool weather and interminable fogs and rains. It has long been cultivated in America, but with little success ; for though frequently abundant, the flavor is indifferent in comparison with American fruits generally. For those who design to cultivate them, the nursery catalogues are a sufficient reference. As a tart, they are inferior to the *rhubarb* or *pie-plant*, which can be grown with little trouble or expense, in great profusion in every fertile and well-tilled garden ; and this is in season from May till August, when apples are sufficiently advanced to take its place.

THE RASPBERRY.

Both the Red and Black Raspberries are favorably known as a wild American fruit. As *a market fruit near the large cities*, it is very profitable. It prefers a light, warm, dry soil, rich and thoroughly loosened. The best varieties grown are the Red and Yellow Antwerps, which produce abundantly, and are of fine flavor; the Franconia, a fine, large, purple fruit ; and the Fastolf, a late English Red variety of superior size and flavor. The above kinds are all hardy in latitude 43° north. They are propagated by suckers, and should be planted three feet apart if in hills, and four feet if in rows. The stalk lives but two years. The first season it shoots up from the root, and makes its growth. The next spring it should be topped to three feet in height, the old stock cut out, and the bearing ones (which ought never to exceed three or four in a clump), should be securely tied to a stake or trellis. If the ground be well hoed, they will bear profusely.

THE STRAWBERRY.

This delicious and wholesome fruit is rapidly spreading in garden cultivation throughout the United States. It will flourish in almost any good soil which is not too cold or wet. The plants should be set in rows two feet asunder, and one foot apart in the rows, kept clear from weeds, and the runners cut off once or twice in the growing season. Beds will last from three to six years, depending in a measure, on the mode of cultivation. The fruit is in season from three to six weeks, according to their kinds. Many horticulturists have found difficulty in procuring an abundant supply of the strawberry, which is probably owing (when other circumstances are favorable), to an improper arrangement of the male and female plants. Hovey's Seedling and several others demand the presence of the male plant from some other variety, to fertilize them The most popular for the market are sub-varieties of the Scarlet, Pine, Chili, and Wood. The Methven Castle, Keene's, Hovey's Seedlings and Boston Pine are among the most highly celebrated.

THE AMERICAN CRANBERRY (Oxycocus macrocarpus)

Yields one of the most delicious of our tarts. It is found in great abundance in many low, swampy grounds in our northern and southern States; and although it has been gathered from its native haunts from the earliest settlement of the country, yet it is only within a few years that it has become an object of cultivation. Experience has probably not yet fully developed the most certain means of attaining the greatest success, but enough is already known, to assume that they are a profitable object of attention to the farmer.

There seems to be several varieties of the cranberry, which differ in size, color, shape and flavor. Some of these are worth much more in the market than others; and, occasionally, the choicest have sold as high as $3.50 per bushel.

Soil and cultivation.—They are generally planted on low, moist meadows, which are prepared by thorough plowing and harrowing. They are then set in drills by slips and roots, usually in the spring, but sometimes in autumn, about 20 inches apart, and at distances of about three inches. They require to have the weeds kept out, and the ground stirred with a light cultivator or hoe, and they will soon overrun and occupy the whole ground. An occasional top dressing of swamp muck is beneficial. In this way, 300 bushels per acre have been produced in Massachusetts, which were worth in the market from one to two dollars per bushel. Capt. Hall, of the same State, raises them in a swamp, first giving it a top dressing of sand or gravel to kill the grass, when he digs holes four feet apart, and inserts in each, a sod of cranberry plants about one foot square. From these sods they gradually spread till the whole surface is occupied.

The cranberry is sometimes killed by late or early frosts; and it has been suggested, that these can be avoided by having the fields so arranged, when frosts may be expected, as to be slightly covered with water. The cranberry is gathered when sufficiently ripe, by raking them from the bushes. They are cleaned from the stems, leaves, and imperfect berries, by washing and rolling them over smooth boards set on an inclined plane, in the same manner as imperfect shot are assorted. After this, they are put into tight casks and filled with water. If stored in a cool place, the water changed at proper intervals, and the imperfect berries occasionally thrown out, they will keep till the following summer. They will frequently bring $20 per barrel in European markets. The raking of the plant in harvesting is beneficial rather than otherwise; for though some of the plants are pulled out and others broken, their places are more than supplied by the subsequent growth.

CHAPTER XII.

MISCELLANEOUS AIDS AND OBJECTS OF AGRICULTURE.

I HAVE thus far treated of soils and manures, the preparation of the ground, and the ordinary cultivated field crops, as fully as my limits permit. It remains briefly to add such incidental aids and objects of agriculture, as could not appropriately be embraced under either of the foregoing heads.

ROTATION OF CROPS—ITS USES AND EFFECTS.

The practice of rotation in crops is an agricultural improvement of very modern date. It is first mentioned in Dickson's Treatise on Agriculture, published in Edinburgh, in 1777. For more than a century it has been partially practiced in Flanders, and perhaps in some other adjoining and highly-cultivated countries. It was afterwards introduced, and imperfectly carried out on a limited scale in the Norfolk district in Great Britain; but its general introduction did not take place till the beginning of the present century. The system of rotation is one of the first and most important principles of general husbandry, and it cannot be omitted without manifest disadvantage and loss. Its place was formerly supplied by *naked fallows*. This practice consists, as I have before shown, in giving the soil an occasional or periodical *rest*, in which no crop is taken off, and the soil is allowed to produce just what it pleases or nothing at all, for one or more years, when it is refreshed and invigorated for the production of its accustomed useful crops. This system, it will be perceived, implies the loss of the income from the soil for a certain portion of the time, and it can be tolerated only where there is more land than can be cultivated.

Modern agricultural science has detected, in part at least, the true theory of the necessity for rotation. It has been discovered, that every crop robs the soil of a part of its elements, (fifteen or sixteen elementary substances combined in various forms and proportions), and that no two dissimilar crops abstract these elements or their compounds from the soil, in the same proportions. Thus, if we consider the amount of the salts taken out of the soil by a crop of turneps, amounting to five tons of roots per acre; of barley, 38 bushels; one ton each of dry clover or rye grass; and of wheat, 25 bushels, we shall find the great disproportions of the various elements, which the different vegetables have appropriated. As given by Johnston they will be in pounds as follows:

	Turnep Roots.	BARLEY.		Red Clover.	Rye Grass.	WHEAT.		Total.
		Grain.	Straw.			Grain.	Straw.	
Potash	145.5	5.6	4.5	45.0	28.5	3.3	0.6	233.0
Soda	64.3	5.8	1.1	12.0	9.0	3.5	0.9	96.6
Lime	45.8	2.1	12.9	63.0	16.5	1.5	7.2	149.0
Magnesia	15.5	3.6	1.8	7.5	2.0	1.5	1.0	32.9
Alumina	2.2	0 5	3.4	0.3	0.8	0.4	2.7	10.3
Silica	23.6	23.6	90.0	8.0	62.0	6.0	86.0	299.2
Sulphuric Acid.	49.0	1.2	2.8	10.0	8 0	0.8	1.0	72.8
Phosphoric do..	22.4	4 2	3.7	15.0	0.6	0.6	5.0	51.5
Chlorine	14.5	0.4	1.5	8.0	0.1	0.2	0.9	25.6
								970 9*

Besides the elements above noted, all crops absorb oxide of iron, and nearly all oxide of manganese and iodine; and of the organic elements associated in various combinations, they appropriate about 97 per cent. of their entire dried weight. Now, it is not only necessary that all the above materials exist in the soil, *but that they are also to be found in a form precisely adapted to the wants of the growing plant.* That they exist in every soil, in some condition, to an amount large enough to afford the quantity required by the crop, can hardly be doubted; but that they are all in a form to supply the full demands of a luxuriant crop, is probably true of such only as are found, under favorable circumstances of season and climate, to have produced the largest burthens.

If a succession of any given crops are gathered and carried off the land, without the occasional addition of manures, they will be found gradually to diminish in quantity, till they reach a point when they will scarcely pay the expenses of cultivation. I mean to be understood as affirming this of all crops and all soils, however naturally fertile

* This is exclusive of the turnep tops.

the latter may be ; unless they are such as receive an annual or occasional dressing from the overflow of enriching floods, or are artificially irrigated with such water as holds the necessary fertilizing matters in solution; and such are not exceptions, but receive their manure in another form, unaided by the hand of the husbandman. Neither are *old meadows* (mowing lands filled with the natural or uncultivated grasses, or whatever of useful forage they choose to bear), exceptions to this rule; although they may part with a portion of their annual crop in the hay, which is removed, and which is not returned as manure, and by a partial rest or pasturage, appear to sustain their original fertility. But if the true character of the various plants which they produce, were accurately observed, and all of which are indiscriminately embraced under the general head of grass or hay, it would be found that the plants gradually change from year to year ; and while some predominate in one season, others take their place the year succeeding, and these again are supplanted by others in an unceasing round of natural rotation.

Another illustration of rotation may be observed in the succession of forest trees that shoot up on the same soil, to supply the places of such of their predecessors as have decayed or been cut down. Thus, the pine and other of the coniferæ, are frequently found to usurp the place of the oak, chestnut, and other deciduous trees. This sometimes occurs only partially, but in repeated instances which have come within my notice, forests have been observed to pass entirely from one order of the vegetable creation to its remote opposite, the seeds or germs of which (the product of an ancient rotation), had been lying dormant for centuries, perhaps, waiting a favorable condition of circumstances and soil to spring into life.

Many choice secondary bottom lands, and others munificently supplied by nature with all the materials of fertility, have, by a long succession of crops, been reduced to a condition of comparative sterility. Yet it will have been found in the progress of this exhaustion, that after the soil ceased to give an adequate return of one crop, as of wheat, corn, or tobacco, it would still yield largely of some other genus which was adapted to it. These lands, when thus reduced, and turned out to commons for a few years, will again give crops much larger than those which closed their former bearing career, proving that nature has been silently at work in renovating the land for further use. The whole course of her operations is not yet known; but it is satisfactorily ascertained, that she is incessantly engaged in producing those changes in the soil, which enable it to contribute to vegetable sustenance. Enough of lime, or potash, or silica may have been disengaged to yield all that may be required for one crop, which by that crop is principally taken up, and if another of the same kind follows in quick succession, there will be a deficiency; yet if a different crop succeed, there may be found enough of all the materials it needs, fully to mature it. A third now takes its place, demanding materials for nutrition, in forms and proportions unlike either which has preceded it, and by the time a recurrence to the first is necessary, the soil may be in a condition again to yield a remunerating return. These remarks apply equally to such soils as have, and such as have not received manures; unless, as is seldom the case, an accurate science should add them in quantity and character (*specific manures*), fully to supply the exhaustion. The addition or withholding of manures, only accelerates or retards this effect.

Another prominent advantage of rotation, is in its enabling such crops to have the benefit of manure, as cannot receive it without hazard or injury if applied directly upon them. Thus wheat and the other white grains, are liable to overgrowth of straw, rust, and mildew, if manured with recent dung; yet this is applied without risk to corn, roots and most of the hoed crops; and when tempered by one season's exhaustion, and the various changes and combinations which are effected in the soil, it safely ministers in profusion to all the wants of the smaller cereal grains. An additional benefit of rotation is, by bringing the land into hoed crops at proper intervals, it is cleared of any troublesome weeds which may infest it. And a still further advantage may be found, in cutting off the appropriate food of insects and worms, which, in the course of time, by having a full supply of their necessary aliment, and especially if undisturbed in their haunts, will ofttimes become so numerous as seriously to interfere with the labors of the farmer. A variation of the crop, exposure of the insects to the frosts, and the change of cultivation which a rotation insures, will make serious inroads upon their numbers, if it does not effectually destroy them.

The noxious excretions of plants, a fanciful theory first

broached and ingeniously defended by the powerful name of Decandolle, and which the closest scrutiny of scientific observers since, has pronounced unworthy of credit, does not form another reason for rotation   It is because principles essential to successful vegetation have been abstracted, not that others hurtful to it have been added to the soil by preceding crops, which renders rotation necessary.

From all that has hitherto been learned on the subject of rotation, either from science or practice, two general principles may be assumed as proper to guide every farmer in his course of cropping.   First, to cultivate as great a variety of plants as his soil, circumstances and market will justify ; and second, to have the same or any similar species follow each other at intervals as remote as may be consistent with his interests.   From the foregoing observations on the subject, it is evident that the proper system of rotation for any farmer to adopt, must depend on all the conditions by which he is surrounded, and that it should vary according to these varying circumstances.

It is a practice with some to alternate wheat and clover, giving only one year to the former and one or two years to the latter.   This will answer for a long time on soils adapted to each crop, provided there be added to the clover, such manures as contribute to its own growth, and such also, as are exhausted by wheat.   The saline manures, ashes, lime, &c., may be added directly to the wheat without injury ; but gypsum should be sown upon the clover, as its benefits are scarcely perceptible on wheat, while upon clover, they are of the greatest utility.   But there are objections to this limited variety, as it does not allow an economical or advantageous use of barn-yard manures, which, from their combining all the elements of fertility, are the most certain in their general effect.   In different countries of Europe, fields which have been used for an oft-recurring clover crop, have become *clover-sick*, as it is familiarly termed.   The plant will not grow luxuriantly, sometimes refusing to vegetate, or if it starts upon its vegetable existence, it does so apparently with the greatest reluctance and suffering, and ekes out a puny, thriftless career, unattended with a single advantage to its owner.   This is simply the result of the exhaustion of one or more of the indispensable elements of the plant.   It it be desirable to pursue this two-course system for any length of time, *nothing short of the application of all such inorganic matters as are taken up by the crops, will sus-*

12*

tain the land in a fertile condition. I subjoin, simply
for the purpose of illustration, and the guidance of such persons as may have little experience in rotation, some sys-
tems which have been pursued with advantage in this country.

1°. On a grass sod broken up, with a heavy dressing of
barn-yard manure, or muck, ashes, and lime if necessary.
First year, corn with gypsum scattered over the plants after
the first hoeing, which should be done immediately upon its
making its appearance; second year, roots with manure;
third year, wheat if adapted to the soil, with guano; if not
then barley, rye or oats, with grass or clover seed or both;
fourth year, meadow, which may be continued at pleasure, or
till the grass or clover gives way. The meadows may be
followed by pasturing if desired. Clover alone should not
remain over two years as meadow, but for pasture it may be
continued longer.

2°. First year, corn or roots on a grass or clover ley with
manure; second, oats and clover, with a top dressing of 10 to
20 bushels of crushed bones per acre; third, clover pastured
to last of June, then grown until fully matured in August,
when it is turned over, and a light dressing of compost and
40 to 80 bushels of leached ashes spread over it, and wheat
and Timothy seed sown about 15th September. If desired,
clover is sown the following spring. This gives for the
fourth year, wheat; fifth and sixth, and if the grass continues
good, the seventh year also, meadow.

3°. First, corn on a grass sod heavily manured, and a half
gill of ashes and gypsum mixed at the rate of two of the
former to one of the latter, and put in the hill, and a less
quantity of pure gypsum added after the corn is first hoed
second, oats or barley, with lime at the rate of 20 or 30
bushels per acre, sown broadcast after the oats and harrowed
in; third, peas or beans, removed early, and afterwards sown
with wheat; fourth, wheat with a light top-dressing of compost, guano and saline manures in the spring, and clover, or
grass and clover seed; fifth, two or three years in meadow
and pasture.

4°. First, wheat on a grass sod; second, clover; third,
Indian corn, heavily manured; fourth, barley or oats, with
grass or clover seed; fifth, and following, grass or clover,
with guano.

5°. A good rotation for light, sandy lands, is first, corn
well manured and cut off early and removed from the ground,
which is immediately sown with rye, or the rye hoed in be-

tween the hills second, rye with clover sown in the spring, and gypsum added when fairly up; third, clover cut for hay, or pastured, the latter being much more advantageous for the land.

WEEDS.

Whatever plants infest the farmer's grounds, and are worthless as objects of cultivation, are embraced under the general name of weeds. In a more comprehensive sense, all plants, however useful they may be as distinct or separate objects of attention, when scattered through a crop of other useful plants to their manifest detriment, may be considered and treated as weeds. Perfect cultivation consists, in having nothing upon the ground but what is intended for the benefit of the farmer; and it implies a total destruction of every species of vegetation, which does not contribute directly to his advantage.

In China and some parts of Flanders, the fields are entirely free from weeds. This is the result of long-continued, cleanly cultivation, by which every weed has been extirpated; a scrupulous attention to the purity of the seed, and the sole use of urine, poudrette and saline manures. This object is scarcely attainable in this country, except on fields peculiarly situated. The principal causes of the propagation of weeds among us, is the negligent system of tillage and the use of unfermented vegetable manures. By heating or decomposition, all the seeds incorporated in the manure heap are destroyed. But there is a great loss in applying manure thus changed, after having parted with large portions of its active, nutritive gases, unless it has been protected by a thick covering of turf or vegetable mold during the progress of fermentation. For many soils and crops, undecomposed manures are far the most valuable. But they should always be applied to the hoed crops, and such as will receive the attention of the farmer for the utter extinction of weeds. A single weed which is allowed to mature, may become 500 the following year, and 10,000 the year after.

The cleansing of land from weeds, is almost the sole justification for naked fallows. When a large crop of them have by any means obtained possession of the ground, they ought to be turned into the soil with the plow before ripening their seed, and they thus become a means of enriching rather than of impoverishing the ground. Meadows which have become foul with useless plants, may be turned into pasture, and if there are plants which cattle and horses will not eat, let them first crop it closely, and then follow with sheep which are much more indiscriminate in their choice of food, and consume many plants which are rejected by other animals. Whatever escapes the maw of sheep, should be extirpated by the hand or hoe before seeding. The utmost care also, should be used in the selection of seed, and none sown but such as has been entirely freed from any foreign seeds.

The Canada thistle is the only weed which has taxed the ingenuity of vigilant farmers in effecting its removal. This is, however, within the power of every one, who will bestow upon it a watchful attention for a single season. The plant should be allowed to attain nearly its full growth, or till it comes fully into flower, when it has drawn largely upon the vitality of its roots. If the bed be large, the plow should be used to turn every particle of the plant under the surface, and let the hoe or spade complete what has escaped the plow. If small or difficult to reach with the plow, use the hoe or spade to cut off the crown of the root ; and if in blossom, let the tops be burnt to prevent the possibility of any of the seeds ripening. As soon as the tops again make their appearance above ground, repeat the plowing or spading. Continue this till the middle of autumn, when the land will be free from them, and in fine condition to yield a crop of wheat. If they harbor in fences or walls, these should either be removed, or the thistle followed to its roots, and kept constantly cut into the ground, when it will not long survive.

An abundance of weeds implies negligence of cultivation and generally, a deficiency of manures is equally conspicuous. The weeds in this case are kindly provided by nature, partially to sustain a fertility which could not otherwise long subsist. When found in any considerable quantity, plow them in before the seed is formed ; and they are frequently equivalent when thus treated, to a good crop of clover or other dressing of green manures.

FIBROUS COVERING OR GURNEYISM

Is the name given to the practice (conspicuously brought into notice recently by Mr. Gurney of England) of covering grass lands with straw or any similar vegetable matter. It has received the sanction of many eminent agriculturists abroad ; and for the purpose of throwing every improvement before our readers which may possibly benefit them, I sub-join the following from an article on the subject, in the British Farmer's Magazine.

"The fact of a remarkable increase of vegetation from fibrous covering, has now been fully confirmed by numerous and careful experiments. In every instance where the relative quantities of grass were cut and weighed, that operated on by this agency showed an increase of six to one over that of other parts of the fields without manure, and of five to one above that where guano, farm-yard manure, wood ashes, or pigs'-house dung had been applied against it. The quantity of hay obtained from the grass was in the same ratio; the mean of the results from different farms, shows that a ton and a half was obtained where Gurneyism had been used, and only from four to five hundred weight where it had not. In many cases the grass was so slight on the parts of the fields not covered, that it could with difficulty be mowed, and in some cases was considered not worth cutting at all.

The question of quantity is indeed settled. The next question, its comparative goodness, seems also determined. Mr. Gurney, at former meetings, gave it as his opinion that the quality was not inferior to that of other grass; this opinion, he said, was founded on botanical observation and careful chemical analysis. In all cases, cattle eat this grass as readily as they do that of ordinary production, and appear to do as well on it. It has moreover been observed that the milk and cream of cows fed on it have both increased in quantity and improved in quality.

Another very interesting and important fact is developed that this action tends to improve the herbage, by favoring the growth of the more valuable kinds of plants In almost every instance it has very much increased the growth of the Dutch clover. In Belgium, and many parts of the midland counties of England, it is the common practice, in order to destroy the couch grass, &c., to manure twice on the green sod with active compost. The result of this practice is to bring up the more valuable grasses, which, being delicate, require the assistance of art to insure their vigorous growth. The same results follow the action of fibrous covering, but in a more rapid manner, and certainly the quality of the herbage is improved. In many parts of fields where the action had been induced, a beautiful floor of grass now appears; while on those parts left uncovered, the grass is very inferior in appearance, having a considerable quan

tity of couch-grass and bent.   There is no doubt, therefore
that the quality of Gurneyized grass will be found in practice equal, if not superior, to that of ordinary growth.

It was thought that the action of fibrous covering was
occasioned by retarding evaporation, and shading the soil
during the unusually dry season.   This, however, is not the
case ; the same proportional increase of vegetation has gone
on since the wet weather set in, and still continues.   Mr.
Gurney stated at the last meeting that he has found fibrous
covering, in a late experiment during the wet weather, had
brought up the eaver and clover in a barley arish, in which
the seeds had failed from the dry season.''   The kind of soil,
and the circumstances attending the application are not
stated, but I infer from the product on the ground, that it
was a very thin and light, and probably a dry soil.

The observation has been frequently made in this country,
that many half-cleared pastures, where the trees and brush
had been prostrated and partially burnt, leaving a heavy
covering of old logs and dead branches, gave a much larger
supply of feed than such as had been entirely cleared.   All
the facts and attending circumstances, however, have not
been given with sufficient particularity to draw any well-
settled conclusions ; yet from the generality of the remark
by observing and careful men, there is undoubtedly some
weight due to it.   The same effect has been often claimed
from certain stony fields, which apparently give much larger
returns than others from which the stones had been removed.

If the results are as have been inferred, after deducting
something for what observation or science may possibly not
yet have detected, I would ascribe them to two causes.   1°.
The gradual decomposition of the vegetable covering or
stone, as either may have occupied the ground, and the direct food which they thus yield to the crop ; and 2°. the
greater and more prolonged deposit of dew, which is going
forward through most of the twenty-four hours of every
day on large portions of the field.   (Does the influence of
the shade and moisture promote an unusual deposite of ammonia, nitric acid, or any of the fertilizing gases ?)   I am
inclined to think nitric acid is thus formed in considerable
quantities, and especially where there is an appreciable
quantity of lime in the soil.   Both M. Longchamp, and Dr.
John Davy assert, " that the presence of azotized matter is
not essential for the generation of nitric acid or nitrous salts,
but that the oxygen and azote of the atmosphere when con-

densed by capillarity, will combine in such proportions as to form nitric acid through the agency of moisture and of neutralizing bases, such as lime, magnesia, potash or soda."— (*Ure*). The condition of the soil is precisely analogous to the artificial nitre beds, deducting their excess of manure and calcareous matter. These exist to some extent in every soil, and it is probable, under similar circumstances they will produce an amount of nitric acid proportionate to their own quantity, which in every case will be particularly felt by the crops. We have the shade, moisture, and capillary condition similar to those of the nitre beds, for the formation and condensing of the acid, which, in this instance, is washed down into the soil by every successive rain, instead of being carefully preserved, where formed, as is done by the roofing of the beds. The question is one of sufficient consequence to induce further trials, under such circumstances as will be likely to afford data for estimating the precise force or influence of these several causes and conditions of the soil.

ELECTRO CULTURE.

The application of electricity to growing plants is a subject which has occupied the attention of scientific men for many years, and apparently without arriving at any beneficial result. That it is capable of producing unusually rapid growth when applied to vegetation, we have too many examples to admit of any doubt. A stream of elec tricity from a galvanic battery, directed upon the seeds or roots of plants under a favorable condition, has sometimes produced an amount of vegetable development within a few hours, which would have required as many days or even weeks to produce, in the ordinary course of nature. An egg has been hatched in one fourth the usual period of incubation ; and every dairy maid is aware of the accelerated change in the milk, from the presence of a highly electrical atmosphere. A thunder storm will sour milk in two hours, that would otherwise have remained unchanged for as many days. But after all the efforts hitherto made to secure this agent for the advancement of the farmer's operations, a careful review of the entire results obtained, compels us to acknowledge, that no application of electricity is yet developed, which entitles it to the consideration of practical agriculturists. When we consider, however, the power and almost universal presence and agency of electricity, we must confess our confidence, that the researches of science will hereafter detect some principles of its operation, which may be of immense value to the interests of agriculture.

Electricity is probably the principal, and perhaps the sole agent in producing all chemical changes in inert matter; nor is it improbable, its agency is equally paramount in the changes of vegetable, and to a certain extent also, of animal life. Independent of human agency or control, it forms nitric acid in the atmosphere during thunder showers, which is brought down by the rain, and contributes greatly to the growth of vegetables. It is also efficient in the deposit of dews, and in numberless unseen ways, it silently aids in those beneficent results, which gladden the heart, by fulfilling the hopes of the careful and diligent husbandman. But until something is more definitely established in relation to its principles and effects, the prudent agriculturist may omit any attention to the subject of electro culture.

EXPERIMENTS AMONG FARMERS.

A great advantage would result to agriculture, if every intelligent farmer would pursue some systematic course of experiments, on such a scale, and with such variety as his circumstances justified, and give the results if successful, to the community. It is with experiments in farming, as was said by Franklin, of a young man's owning wild lands; " it is well for every one to have some, *if he don't have too many.*" They should be his servants, not his masters; and if intelligently managed and kept within due bounds, they may be made to subserve his own interest, and by their promulgation, eminently promotive of the general good. It is fully in accordance with another maxim of that wise head, that when it is not within our power to return a favor to our benefactor, it is our duty to confer one on the first necessitous person we meet, and thus the circle of good offices will pass round.

The mutual communication of improvements of any kind in agriculture, has the effect of benefiting not only the community generally, but even the authors themselves; as they frequently elicit corrections and modifications which materially enhance the value of the discovery. These experiments should embrace the whole subject of American agriculture; soils and their amelioration; manures of every kind, alkaline, vegetable and putrescent, and their effects on different soils and crops ; plants of every variety, and their adaptation to different soils, under different circumstances, and with various manures ; and their relations to each other, both as successors in rotation, their value for conversion into animals and other forms, and their comparative ultimate profit ; the production of new varieties by hybridizing and otherwise ; draining, both surface and covered ; the improvement of implements and mechanical operations, &c., &c. They should also extend to the impartial and thorough trial of the different breeds of all domestic animals, making ultimate profit to the owner the sole test of their merits ; crossing them in different ways, and under such general rules as experience has determined as proper to be observed ; their treatment, food, management, &c. Although much has been accomplished within the last few years, the science and practice of agriculture may yet be considered almost in its infancy. There is an unbounded field still open for exploration and research, in which the efforts of persevering genius, may hereafter discover mines of immense value to the human family.

THE UTILITY OF BIRDS.

These are among the most useful of the farmer's aids, in securing his crops from insect depredations ; and yet manifest as this is to every observing man, they are frequently pursued and hunted from the premises as if they were his worst enemies. The martin, the swallow and the wren, which may almost be considered among the domestics of the farm ; and the sparrow, the robin, the blue bird, the wood-pecker, the bob-a-link, the thrush, the oriole, and nearly all the songsters of the field accomplish more for the destruction of noxious flies, worms and insects, which are the real enemies of the farmer, than all the nostrums ever invented. And hence the folly of that absurd custom of scare-crows in the growing corn-fields and orchards, to which I have before alluded. The chickens and ducks do the farmer more benefit than injury in the garden and pleasure grounds, if kept out of the way while the young plants are coming up. A troop of young turkeys in the field, will destroy their weight in grasshoppers every three days, during their prevalence in summer or autumn. A pair of sparrows, while feeding their young, consume over 3,000 caterpillars a week. One hundred crows devour a ton and a half of grubs and insects in a season. Even the hawk and the owl, the objects of general aversion, rid the fields and woods of innumerable squirrels, moles and field mice. The last are frequently great depredators upon the crops, after having exhausted the stores of worms and insects which they first invariably devour, and to this extent these little quadrupeds are themselves benefactors. The smaller species of the hawk and owl, when pressed by hunger, will resort to grubs, beetles, crickets and grasshoppers, in the absence of larger game. That loathsome monster the bat, in its hobgoblin flight, destroys his bulk of flies in a single night. Slight injury may occasionally be done to the grain and fruit by the smaller birds, and when thus intrusive, some temporary precaution will suffice to prevent much loss. But whatever loss may thus occur, the balance of benefit to the farmer from their presence, is generally in their favor; and instead of driving them from his grounds, he should encourage their social, chatty visits by kind and gentle treatment, and by providing trees and pleasant shrubbery for their accommodation.

TOADS, FROGS, &c.

Shakspeare has said,

> The toad, ugly and venomous,
> Wears yet a precious jewel in its head.

Deducting the *venom* we shall find the poet right; for we can no more attempt the defence of his beauty, than that of the muck heap; but we can well excuse his unprepossessing exterior, for the sake of the jewel which he wears in his tongue. This, like that of the chamelion, of which he is a cousin-german, he darts out with lightning rapidity, and clasps the worm or insect prey within its glutinous folds, which, with equal rapidity, is transformed to his capacious maw. Apparently dull, squat, and of the soil's hue, whatever that may be; he sits quiet and meditative, yet watchful in the thick shade of some overgrown cabbage; and as the careless insects buzz by, or the grub or beetle crawl lazily along, unheedful of danger, he loads his aldermanic carcass with the savory repast. Sixteen fresh beetles, a pile equal to his fasting bulk, have been found in the stomach of a single toad.

The frog, traipsing over the wet fields, amid the long grass or thick weeds, procures his summer subsistence in the same way as his seeming congener the toad, and with equal benefit to the farmer.

The striped snake is a harmless object about the farm premises, and like the toad, he is also a great gormandiser of worms and insects. The sole drawback to his merits, arises from his frequently feasting on the toad and frog. *The black snake* is sometimes destructive to young poultry, and he is a fierce and formidable foe to all whom his courage induces him to attack. He charms the old birds and robs their nests both of eggs and young; but his consumption of superfluous squirrels and field mice, perhaps, fully atones for his own delinquencies.

FENCES.

In many countries which have long been under cultivation, with a dense population and little timber, as in China and other parts of Asia, Italy, France, Belgium, Holland and other parts of Europe, fences are seldom seen. In certain sections of the older settled portions of the New England States, a similar arrangement prevails. This is universally the case over the wide intervals or bottom lands which skirt the banks of the Connecticut River, where periodical inundations would annually sweep them away. Wherever this system is adopted, cultivation proceeds without obstruction, and a great saving is made not only in their original cost, but in the interest, repairs and renewal; all the land is available for crops; no weeds or bushes are permitted to hide their annoying roots and scatter their seeds over the ground; no secure harbors are made for mice, rats or other vermin; the trouble and expense of keeping up bars or gates are avoided; and a free course is allowed by the conceded roads or by-paths, for the removal of the crops, carrying on manures, and the necessary passing to and fro in their cultivation. These are important advantages, which it would be well for every community to consider, and secure to the full extent of their circumstances.

The inconveniences of this arrangement are trifling. When cattle or sheep are pastured where fences are wanting, they are placed under the guidance of a shepherd, who with the aid of a well-trained dog, will keep a large herd of animals in perfect subjection within the prescribed limits. In the unfenced parts of the Connecticut valley (where extensive legislative powers reside in the separate towns, which enables each to adopt such regulations as best comport with their own interests), no animals are permitted to go upon the unfenced fields till autumn; and the crops are required to be removed at a designated time, when each occupant is at liberty to turn upon the common premises, a number of cattle proportionate to his standing forage, which is accurately ascertained by a supervisory board. A certain number of fences are necessary for such fields as are continued in pasture through the season, but unfortunately, custom in this country has increased them beyond all necessity or reason. It rests with the farmers to abate such as they deem consistent with their interests.

The kind of fences required, must vary according to the controlling circumstances of the farm. In those situations where stone abounds, and especially if it is a nuisance, heavy stone fences (broad and high) are undoubtedly the most proper. Where these are not abundant, an economical fence may be constructed, by a substantial foundation of stone, reaching two or two and a half feet above ground, in which posts are placed at proper distances, with two or three bar holes above the wall, in which an equal number of rails are inserted. Post and rail, and post and board fences are common where there is not a redundancy of timber.

The posts should be placed from two and a half to three feet below the surface, in the centre of a large hole and surrounded by fine stone, which must be well pounded down by a heavy, iron-shod rammer, as they are filled in. The post will not stand as firmly at first as if surrounded by dirt, but it will last much longer. The lower end should be pointed, which prevents its heaving with the frost. If the position of the post while in the tree be reversed, or the upper end of the split section of the trunk which is used for a post, be placed in the earth, it will be more durable. Charring or partially burning the part of the post which is buried, will add to its duration. So also will imbedding it in ashes, lime, charcoal or clay: or it may be bored at the surface with a large auger, diagonally downwards and nearly through, then filled with salt, and closely plugged.

The best timber for posts in the order of its durability, is red cedar, yellow locust, white oak and chestnut, for the northern and middle States. I recently saw red cedar posts in use for a porch which, I was assured had been standing exposed to the weather previous to the Revolution, a period of over 70 years, and they were still perfectly sound. The avidity with which silicious sands and gravel act upon wood, renders a post fence expensive for such soils.

There are large portions of our country where timber

abounds, especially in the uncleared parts of it, where the zig-zag or worm fence is by far the most economical. The timber is an incumbrance and therefore costs nothing; and the rails can be cut and split for 50 to 75 cents per 100, and the hauling and placing is still less. With good rails, well laid up from the ground on stones or durable blocks, and properly crossed at the ends and locked at the top, these fences are firm and durable.

Staking the corners by projecting rails gives an unsightly appearance at all times, and is particularly objectionable for plowing, as it considerably increases the waste ground. The same object is obtained by locking the fence when completed, with a long rail on each side, one end resting on the ground and the other laid into the angle, in a line with the fence. More symmetry and neatness is secured, and a trifling amount of timber saved, by putting two small upright stakes, one on each side of the angle, and securing them by a white oak plank, six inches wide by eighteen inches long, with holes to slip over the posts, after most or all of the rails have been laid. Any additional ones laid over this, keep the yokes or caps in their place, and the whole is thus firmly bound together. Besides the timber designated for posts, rails may be made from any kind of oak, black walnut, black and white ash, elm and hickory.

Turf and clay fences have been tried in this country without success. The frosts and rains are so severe as to break and crumble them down continually. Cattle tread upon and gore them; and against swine and sheep they scarcely offer any resistance.

Wire fences have been tried successfully. They are made with a greater or less number of wires and of sizes

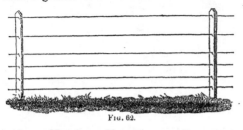

Fig. 62.

varying from No. 6 to No. 12, according to the style of building and the purposes to be answered. If wanted for the larger animals, the wires should be stronger, and placed higher; if for the lesser, they may be of smaller wires, run nearer together and closer to the ground.

Fig. 62 shows one of the plain fences, secured by iron or wooden posts, as may be preferred. Fig. 63 is a wire fence, more elegant in design, but much more expensive. In this, the upright wires are secured by longitudinal iron bars. It is equally efficient with the former, and much more ornamental, and is a pretty appendage about the pleasure grounds. The wire may be prepared against rust from the elements, by galvanizing, or they ·may be painted after being put up.

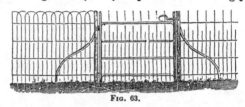

Fig. 63.

Where wood abounds, rails may perhaps, make the most economical fences. But where there is a deficiency, wire is undoubtedly the cheapest. They possess, also, the further advantage of much greater durability; and the facility for removal, at little trouble or expense whenever required.

The hurdle or light moveable fence is variously formed of cordage, wire or wicker work, in short panels, and firmly set in the ground by sharpened stakes at the end of each panel, and these are fastened together. This is a convenient addition to farms where heavy green crops of clover, lucern, peas or turneps are required to be fed off in successive lots, by sheep, swine or cattle.

The sunken fence or wall is by far the most agreeable to good taste, and it is perfectly efficient. It consists of a vertical excavation on one side, about five feet in depth, against which a wall is built to the surface of the ground. The opposite side is inclined at such an angle as will preserve the sod against sliding, from the effects of frost or rain, and is then turfed over. A farm thus divided, presents no obstruction to the view, while it is everywhere properly walled in, besides affording good ditches for the drainage of water. These sunken fences are sometimes raised a couple of feet above the ground, which increases the protection, and at a less cost than deepening and widening the ditch.

Good fences, at all times kept in perfect repair, are the cheapest. Most of the unruly animals are taught their habits through the negligence of their owners. Fences that are half down or which will easily fall by the rubbing of cattle, will soon teach them to jump, and throw down such as they are unable to overleap. For the same reason, gates are better than bars. When the last are used, they should be let down so near the ground that every animal can step over conveniently; nor should they be hurried over so fast as to induce any animal to jump. In driving a flock of sheep through them, the lower bars ought to be taken entirely out, or they be allowed to go over the bars in single file. Animals will seldom become jumpers except through their owner's fault, or from some bad example set them by unruly associates; and unless the fences be perfectly secure, such ought to be stalled till they can be disposed of. The farmer will find that no animal will repay him the trouble and cost of expensive fences and ruined crops.

HEDGES.

These have, from time immemorial, been used in Great Britain and some parts of the European continent, but are now growing unpopular with utilitarian agriculturists. They occupy a great deal of ground, and harbor much vermin. A few only have been introduced in this country, and they will probably never become favorites among us. For those disposed to try them as a matter of taste or fancy, I enumerate as best suited to this object, the *English hawthorn*, beautiful and hardy; the *holly*, with an evergreen handsomely variegated with yellow spots, and armed at the edges with short stiff thorns; the *gorse* or *furze*, a prickly shrub growing to the height of five feet or more and bearing a yellow blossom. These are much cultivated in Europe as defences against the inroads of animals; while numerous other less formidable shrubs, like *the willow* and *privet* are grown for protection against winds; and when sufficiently large and strong, they also serve for cattle enclosures.

Buckthorn Hedge (Fig. 64).

In America the *buckthorn* was first introduced by Mr

Derby, of Massachusetts, and by him was considerably disseminated through the United States. It has proved a hardy, thrifty plant, entirely suited to the purpose. The foregoing figure shows the thorn hedge, which is impervious to any intruder when properly trained. The *Osage orange* grows spontaneously in the southwestern States, and is successfully cultivated in most of the eastern and portions of the northern. Its rapid growth and numerous, thick, tough branches and thorns, render it an effectual protection to fields.

The Cherokee rose, for the southern States, is by many deemed the most economical and efficient. It grows with great rapidity, some of the runners reaching forty feet or more, and it is hardy and enduring; but it occupies much room and is frequently troublesome about the fields from its superabundant growth. *The wild peach*, which abounds in many of the southern forests, furnishes one of the most beautiful hedges when tastefully managed, as I have seen it in Mississippi. I have fenced with the *native thorn* of western New York, with entire success. The *Michigan rose* and the *sweet briar*, both hardy and of luxuriant growth, and some other species of the native rose, have been tried and proved efficient. The *crab apple* and *wild plum*, with their thick, tough branches and formidable thorns (especially the latter), with proper training, will be found a perfect stoppage against animals of all kinds. The *yellow locust* and *acacia* have been sometimes used; and the *wild laurel*, an evergreen of great beauty at all times, and especially so with its magnificent blossoms, would form a beautiful hedge wherever the soil will give it luxuriant growth. There are a variety of other trees and shrubs of native growth, which may be employed for hedges; but it is unnecessary to specify them, as each can best select for himself such as are suited to his own peculiar soil and circumstances.

SHADE TREES.

In such situations and numbers as may be required around the farm premises, these are both ornamental and profitable. They have, too, a social and moral influence, far beyond the mere gratification of the eye or the consideration of dollars and cents. In their freshness and simplicity, they impress the young mind with sentiments of purity and loveliness as enduring as life. From the cradle of infancy, consciousness first dawns upon the beauty of nature beneath their grateful shade; the more boisterous sports of childhood seek their keenest enjoyment amid their expanding foliage ; and they become the favorite trysting place when the feelings assume a graver hue, and the sentiments of approaching manhood usurp the place of unthinking frolic. Their memory in after life greets the lonely wanderer amid his trials and vicissitudes, inciting him to breast adversity till again welcomed to their smiling presence. Their thousand associations repress the unhallowed aspirations of ambition and vice; and when the last sun of decrepid age is sinking to its rest, these venerable monitors solace the expiring soul with the assurance, that a returning spring shall renew its existence beyond the winter of the tomb.

Trees ought not to be too near the buildings, but occupy such a position as to give beauty and finish to the landscape In addition to danger from lightning, blowing down, or the breaking off of heavy branches, there is an excessive dampness from their proximity, which produces rapid decay in such as are of wood, besides its frequently affecting to a serious degree the health of the inmates. Low shrubbery that does not cluster too thickly, or immediately around the house, is not objectionable. Trees are ornamental to the streets and highways, but should be at such a distance from the fences, as will prevent injury to the crops and afford a kindly shade to the wayfarer. In certain sections of the middle and southern States, where the soil is parched from the long sultry summers, it has been found that shade trees rather increases than diminishes the forage of the pastures ; but through most of the middle and northern States, they are decidedly disadvantageous, as the feed is found to be sweeter and more abundant beyond their reach. For this reason, such trees as are preserved exclusively for timber should be kept together in the wood-lots, and even many that are designed for necessary shade or ornament, may be grouped in tasteful copses, with greater economy of ground and manifest improvement to the landscape. In the selection of trees, regard should be had not only to the beauty of the tree and its fitness for shade, but to its ultimate value as timber and fuel.

In the range of selection, no flora of either hemisphere will compare with the number, variety, and beauty of our North American forest trees. Of the oak, we have 50 species, while all Europe has but 30. Europe has 14 species of pines and firs ; Asia, 19 ; Africa, South America, and

13

Polynesia—each, two ; Australia, one ; wh le North America has 40, and the United States alone ove: 20.

The Oak, of which Fig. 65 affords a splen.did specimen, is

one of the most magnificent, as it is one of the longest-lived of the forest tribes. It is to be regretted that many of the species are so slow of growth, that they seldom tempt their cultivation by the utilitarian of this country, who looks only to the speedy enjoyment of his labor. Many of those which have descended to us from pre-

FIG. 65.

ceding generations, combine much of the beauty, and all of the grandeur we can expect in a shade tree.

The Black Oak (Fig. 66), on soil adapted to it, is a tree of commanding beauty and stalwart growth. The foliage appears late, but is unsurpassed for depth and richness of color, and highly-polished surface ; and it retains its summer green, long after the early frosts have mottled the ash, and streaked the maple with their rainbow hues. When grown on dry and open land, both fuel and timber are firm, compact and lasting.

FIG. 66.

The Pin Oak (Figs. 67 and 68), of which two speci-

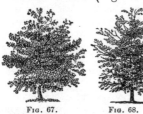

mens are given, grows rapidly and with great beauty, when transplanted into an open space. It affords a timber of great strength and excellence Some others of our nor-:hern oaks yield a fine shade, and good timber

FIG. 67.   FIG. 68.

and fuel when grown in open grounds.

The Live Oak (Fig. 69) often flourishes as far as 35°. north ; but is not found in perfection north of Louisiana. It grows rap.dly, and assumes a great variety of shapes in its earlier growth, but most frequently appears as in Fig. 69 Sometimes, though rarely, it branches stiffly upwards like the ash ; occasionally t is seen dipping its long, drooping boughs into the water, some feet belcw the surface of the bank which supports its roots ; and

FIG. 69.

more often it imitates the maple. But it most frequently throws its branches out abruptly, and nearly at right angles with the trunk, like the apple; or gracefully arches upward till the flexile twigs descend from the outer extremity, forming a beautiful and usually flattened dome, with a diameter sometimes exceeding 150 feet. It renews its foliage generally once in two years, and then gradually, thus always affording a dense mass of living green. Many other species of evergreen oaks, which never grow north of about 37° within the States, do not vary materially in appearance or character from the live oak. For ship-building, the live oak is esteemed the strongest and most durable timber.

The Elm when standing isolated, is one of our most beautiful and imposing trees. It grows to an immense size, with gracefully projecting limbs and long pendant branches. It is liable to few diseases, and the fuel and timber are good for most purposes. Every one who has seen the patriarchal elms which grace the beautiful villages of the Connecticut valley, and other old towns of New England, must wish to see them universally disseminated.

The *Rock or Sugar Maple*, before mentioned on page 219, (Fig. 70), has a straight trunk and regular upward-branching limbs, forming a top of great symmetry and elegance. Besides the ornament and thick shade it affords, it gives an annual return in its sap, which is converted into syrup and sugar. The fuel is not inferior to any of our native trees, the timber is valuable, yielding the beautiful glossy *bird's eye maple* so much esteemed for furniture, and various other purposes.

FIG. 70.

The *Black Walnut* (Fig. 71) is a stately, graceful tree yielding excellent wood and durable timber; and besides its extensive use for plain, susbtantial furniture, the knots and crotches make the rich, dark veneering, which rivals the mahogany or rose wood in brilliancy and lasting beauty. In a fertile soil, where only it is found in its native state, it bears a rich, highly flavored nut.

FIG. 71.

The White Ash has a more slender and stiffer top than either of the preceding, yet is light and graceful. The fuel is good, and the timber unequalled in value for the carriage maker.

There are two species of willow usually cultivated as shade trees. The *White Willow* (Fig.

72, which is rather a superannuated specimen) generally occupies a low, moist situation, on the brink of some rivulet or stream. It is nearly valueless except in the shade it affords. The bright orange twigs and branches furnish an unfailing supply of primitive whistles for the youngsters in the spring. The

FIG. 72.

Weeping Willow is a tree of variegated foliage, and long flexile twigs, sometimes trailing the ground for yards in length. Its soft, silvery leaves are among the earliest of spring, and the last to maintain their verdure in autumn. But its wood is of little value.

The *Locust* (Fig. 73) is a beautiful tree, of rapid growth, flowering profusely, with its layers or massive flakes of innumerable leaflets of the deepest verdure. The wood is unrivalled for durability as ship timber, except by the *live-oak ;* and for posts or exposure to the weather, it is exceeded only by the savin or red cedar. It has of late years, been subject to severe attacks and

FIG. 73.

great injury from the borer, a worm against whose ravages hitherto, there has been no successful remedy.

The *Button-wood, Sycamore, Plane-tree or Water-beach,* by all of which names it is known in different parts of this country, is of gigantic dimensions when occupying a rich and moist, alluvial soil. One found on the banks of the Ohio measured 47 feet in circumference, at a height of four feet from the ground. Its lofty mottled trunk, its huge irregular limbs, and its numerous pendant balls (in which are compressed myriads of seeds with their plumy tufts, that are wafted to immense distances for propagation), have rendered it occasionally a favorite. They are often seen on the banks of our rivers, almost constituting a hedge ; and sometimes they completely span streams of considerable size. The wood is cross-grained, and intractable for working, and the timber is of little use except for fuel.

The *Magnolia* (*Magnolia grandiflora*) is a splendid southern evergreen, with a beautiful fir or cone-like top, bearing leaves greatly increased in size and thickness beyond those of the evergreen oaks, and of equally deep, perennial verdure. Among these, the large snow white blossoms, six or seven inches in diameter and of great fragrance, spangle, in leisure ly succession, the whole circumference, during most of the months of May and June. Its timber is soft, but useful for some purposes.

FIG. 74.

The Pecan (Fig. 74) is a deciduous, nut-bearing tree, of the walnut tribe, and grows in the same .atitudes as the live-oak and magnolia. It is much taller than either, and somewhat resembles in its growth the rock maple of the North. The nuts are of considerable value as an article of food and export.

FIG. 75.

The Paper Mulberry (Fig. 75) is a handsome shade tree. It is a native of Japan, and was introduced into this country in 1784. Its wood and timber are of little value; but great merit is claimed for its leaves as forage for cattle; for its sap as a substitute for glue; and for its bark as a material for both cloth and paper. It is hardy enough for any part of the United States, south of 42°.

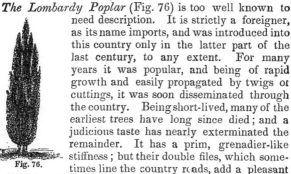
Fig. 76.

The Lombardy Poplar (Fig. 76) is too well known to need description. It is strictly a foreigner, as its name imports, and was introduced into this country only in the latter part of the last century, to any extent. For many years it was popular, and being of rapid growth and easily propagated by twigs or cuttings, it was soon disseminated through the country. Being short-lived, many of the earliest trees have long since died; and a judicious taste has nearly exterminated the remainder. It has a prim, grenadier-like stiffness; but their double files, which sometimes line the country roads, add a pleasant feature to the distant landscape. Its wood and timber are almost worthless, being light, porous and unsubstantial.

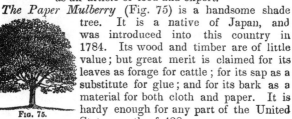

The Tulip Tree (Fig. 77) is one of our most beautiful native shade trees. It abounds on this continent, south of latitude 43°, and a little distance from the Atlantic coast. It grows lofty and large, with a symmetrical top, of great beauty. It has a smooth polished trunk when young, which never becomes very rough or jagged by age. It bears a profusion of delicate, greenish colored, nearly inodorous flowers, as large and not unlike the outline of the cotton blossom. Its timber is light and soft, but useful for many purposes.

THE AILANTUS (*A. glandulosa*, Fig. 78) has not been ex-

FIG. 78.

tensively reared in the United States, till within the last few years. It is hardy, and grows rapidly in latitude 42°, and south of it, throwing out its long, pinnated, sumach-like leaf, from 20 to 70 inches in length. It forms a pleasant shade immediately after transplanting, and will continue to grow rapidly and with great beauty till it attains a height of 50 to 70 feet Some even exceed this size. The Chinese, from whom we get it, give to it the imposing name of *the Tree of Heaven*. Its wood is hard and compact, and of a deep, reddish color. It will receive some polish, and retains a slight lustre, sufficient to justify its use for cabinet work.

The European Larch as shown in fig. 79, is of many

FIG. 79.

varieties, and is sometimes used in this country as a shade tree. For this object, the *Redconed* and *Weeping* varieties are esteemed the most ornamental. There are several other varieties indigenous to America, closely resembling the former in all their peculiarities. They are more generally known in this country as the *Hackmatack* or *Tamarack*. They are partial to moist or swampy and cold soils. The timber is among the strongest and most durable. They sometimes grow to the height of 80 or 100 feet, and two to three feet in diameter.

The Cedar of Lebanon (Fig. 80) endeared to the memo-

FIG. 80.

ry of youth by a thousand incidents of biblical and profane history, is a tree of large size, and peculiar in its widely-outspreading branches. It is frequently used as a shade tree in Europe, but seldom in this country. The great value of its timber may be inferred, from the fact that it was almost exclusively used in the building of the temple of Solomon, whose costly materials and elaborate finish has never been equalled before or since. There is one specimen growing at Throg's Neck, N. Y., two feet in diameter, that produces an abundance of cones annually, from which other trees can be grown.

The Hemlock, (Fig. 81), is a native of all the middle and

northern States. It is an evergreen, slightly
resinous; and when growing in an open space,
has a beautifully symmetrical top. It affords
a dense and agreeable shade from its innumerable leaflets. As fuel, it is better than the
white pines, but inferior to the resinous or pitch
pines, and for timber or lumber it is inferior to
either for most purposes.

Fig. 81.

The Balm of Gilead (Fig. 82) is a native

of this continent, and abounds in low, moist situ
ations, among a great variety of other species of
the fir tribe. It grows well when transplanted
to open ground, where it is sufficiently moist, and
sustains a handsome, pyramidal top, of deep ver
dure. When young and thrifty, it has a thick
foliage, but becomes thin and unsightly when
old ; before which period, it should be removed
from the ornamental grounds.

Fig. 82.

The Long-leaved Pine (Fig. 83) frequently
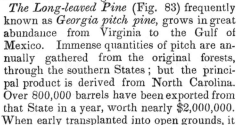
known as *Georgia pitch pine*, grows in great
abundance from Virginia to the Gulf of
Mexico. Immense quantities of pitch are annually gathered from the original forests,
through the southern States ; but the principal product is derived from North Carolina.
Over 800,000 barrels have been exported from
that State in a year, worth nearly $2,000,000.

Fig. 83.

When early transplanted into open grounds, it
forms a fine shade tree. While young, the tufts of leaves
are very long, bushy and graceful.

The White Pine of the North is also a beautiful shade
tree, when not too old. The foliage of all the pines and resinous trees, becomes thin and scraggy as they advance beyond
middle age.

WOOD LANDS.

There are few farms in the United States, where it is not
convenient and profitable to have one or more wood lots attached. They supply the owner with his fuel, which he
can prepare at his leisure; they furnish him with timber for
buildings, rails, posts and for his occasional demands for implements ; they require little attention, and if well managed,
will yield more or less forage for cattle and sheep. The

trees should be kept in a vigorous, growing condition, as the profits are as much enhanced from this cause as any of the cultivated crops.

Few American fields require planting with forest trees. The soil is everywhere adapted to their growth, and being full of seeds and roots, when not too long under cultivation, it needs but to be left unoccupied for a time, and they will everywhere spring up spontaneously. Even the *oak openings* of the West, with here and there a scattered tree, and such of the prairies as border upon woodlands, when rescued from the destructive effects of the annual fires, will rapidly shoot up into vigorous forests. I have repeatedly seen instances of the re-covering of *oak barrens* and *prairies* with young forests, which was undoubtedly their condition before the Indians subjected them to conflagration ; and they have indeed, always maintained their foothold against these desolating fires, wherever there was moisture enough in the soil to arrest their progress.

In almost every instance, if the germs of forest vegetation have not been extinguished in the soil, the wood-lot may be safely left to self-propagation, as it will be certain to produce those trees which are best suited to the present state of the soil. Slightly thinning the young wood may in some cases be desirable, and especially by the removal of such worthless shrubbery as never attains a size or character to render it of any value. Such are the alders, the blue-beach and swamp-willow ; and where there is a redundance of the better varieties of equal vigor, those may be removed that will be worth the least when matured. In most woodlands, however, nature is left to assert her own unaided preferences, growing what and how she pleases, and it must be confessed she is seldom at variance with the owner's interest. Serious and permanent injury has often followed close thinning.

In cutting over woodlands, it is generally best to remove all the large trees on the premises at the same time. This admits a fresh growth on an equal footing, and allows that variety to get the ascendancy to which the soil is best suited. In older settled States, where land and its productions are comparatively high, many adopt the plan of clearing off everything, even burning the old logs and brush, and then sow one or more crops of wheat or rye, for which the land is in admirable condition, from the long accumulation of vegetable matter and the heavy dressing of ashes thus re ceived. They then allow the forest to resume its origina claims, which it is not slow to do, from the abundance of seeds and roots in the ground. But unless the crop be valuable, the utility of this practice is doubtful; as by the destruction of all the young stuff which may be left, there is a certain delay of some years in the after growth of the wood; and the gradual decay of the old trunks and brush, may minister fully as much to its growth as the ash which their combustion leaves; and the fertility of the soil is diminished just in proportion to the amount of vegetable matter abstracted by the grain crops which may have been taken off.

The proper time for cutting over the wood must depend on its character, the soil, and the uses to which it is to be applied. For saw-logs or frame-timber, it should have a thrifty growth of 40 or 50 years; but in the mean time, much scattering fuel may be taken from it, and occasionally such mature timber trees as can be removed without injury to the remainder. For fuel alone, a much earlier cutting has been found most profitable. The Salisbury Iron Company has several thousand acres of land, which have been reserved exclusively for supplying their own charcoal. The intelligent manager informed me, that from an experience of sixty years, they had ascertained the most profitable period for cutting, was once in about sixteen years, when everything was removed of a proper size, and the wood was left entirely to itself for renewed growth. It has been found that this yielded a full equivalent to an annual interest on $16 to $20 an acre, which for a rough and rather indifferent soil, remote from a wood or timber market, will pay fully as much as the nett profits on cultivated land in the same neighborhood.

The wood should be kept entirely free from sheep and cattle, when young, as they feed upon the fresh shoots with nearly the same avidity as they do upon grass or clover; and when it is desirable to thicken the standing trees by an additional growth, cattle should be kept from the range till such time as the new sprouts or seedlings may have attained a height beyond their reach. When it is necessary to bring into woodland such fields as have not forest roots or seeds already deposited in a condition for germination, the fields should be sown or planted with all the different nuts or seeds adapted to the soil, and which it is advantageous to cultivate.

Transplanting trees for a forest in this corntry, cannot at

13*

present be made to pay, from its large expense ; and if the trees will not grow naturally or by sowing, the land should be continued in pastures or cultivation.   There are some lands so unfitted for tillage by their roughness or texture, as to be much more profitable as woodland.   It is better to retain such in forest, and make from them whatever they can thus produce, rather than by clearing and bringing them into use, to add them to superfluous tillage fields, and become a drain on labor and manures which they indifferently repay.

In clearing lands, when it is desirable to reserve trees for a park or shade, a selection should be made of such as are young and healthy, which have grown in the most open places, with a short stem and thick top.   It will tend to insure their continued and vigorous growth, if the top and leading branches be shortened.   Large trees will seldom thrive when subjected to the new condition in which they are placed, after the removal of the shade and moisture by which they have been surrounded.   They will generally remain stationary or soon decay ; and the slight foothold they have upon the earth by their roots, which was sufficient for their protected situation while surrounded by other trees, exposes them to destruction from violent gales ; and they do not acquire or attain that beauty of top and symmetry of appearance which should entitle them to preservation. If partialities are to be indulged for any, they should be surrounded by a copse of younger trees, by which they will be in a measure protected.   Young stocks should be left in numbers greater than are required, as many of them will die, and from the remainder, selections can be made of such as will best answer the purpose designed.

THE PROPER TIME FOR CUTTING TIMBER.

Nine tenths of the community think winter the time for this purpose, but the reason assigned *that the sap is then in the roots*, shows its futility, as it is evident to the most superficial observer, that there is nearly the same quantity of sap in the tree at all seasons.   It is less active in winter, and like all other moisture, is congealed during the coldest weather ; yet when not absolutely frozen, circulation is never entirely stopped in the living tree.   Reason would seem to indicate, that the period of the maturity of the leaf, or from the last of June to the first of November, is the season for cutting timber in its perfection.   Certain it is, that we have numerous examples of the timber cut within this period, which has exhibited a durability twice or three times as great as that cut in winter when placed under precisely the same circumstances. After it is felled, it should at once be peeled, drawn from the woods, and elevated from the ground to facilitate drying; and if it is intended to be used under cover, the sooner it is put there the better. Wood designed for fuel, will spend much better when cut within the same periods, and immediately housed; but as this is generally inconvenient, from the labor of the farm being then required for the harvesting of the crops, it may be more economical to cut it whenever there is most leisure.

Preservation of timber.—Various preparations of late years, have been tried for the more effectual preservation of timber, which have proved quite successful, but the expense precludes their adoption for general purposes. These are kyanizing, or the use of carburetted azote (the base of prussic acid); a solution of common salt; the use of corrosive sublimate, (a bi-chloride of mercury); pyrolignite of iron, formed from iron dissolved in pyroligneous acid, which is produced from the distillation of wood, or from the condensed vapor that escapes from wood fuel while burning, and which may be obtained in large quantities from a coal pit where charcoal is made. These will be absorbed by the sap pores and universally disseminated through the body of the tree, by sawing or cutting the trunk partially off while erect, and applying the solution to its base; or it may be cut down, leaving a part of the leafy branches above the point of saturation, and apply the solution to the butt end. The leaves will continue to sustain the natural flow of the sap, which is both upward and downward, by the different conduits or sap vessels, thus distributing the artificial solution throughout the trunk.

Beautiful tints are given to timber which is used for cabinet work, by saturating it with various coloring matters. Although the expense of these preparations may prevent their use for large, cheap structures, yet for all the lighter implements such as wagons, plows, and tools generally, where the cost of the wood is inconsiderable in comparison with that of making, it would be economy to use such timber only as will give the longest duration, though its first price may be ten-fold that of the more perishable material.

FARMING TOOLS.

These should form an important item of the farmer's at

tention, as upon their proper construction depends much of the economy and success with which he can perform his operations. There have been great and important improvements within the past few years, in most implements, which have diminished the expense, while they have greatly improved the mechanical operations of agriculture. I have studiously avoided a reference to any of these, as there are many competitors for similar and about equally meritorious improvements; and in this career of sharp and commendable rivalry, what is the best to-day, may be supplanted by something superior to-morrow. These implements may now be found at the agricultural warehouses, of almost every desirable variety. Of these, the best only should be procured; such as are the most perfect in their principles and of the most durable materials. Cover the wood work with paint or oil, if to be exposed to the weather, and the iron or steel with paint, or a coating of hot tar, unless kept brightened by use. When required for cutting, they should always be sharp, even to the hoe, the spade and the share and coulter of the plow. When not in use, keep them in a dry place. Plows, harrows, carts and sleds should all be thus protected, and by their longer durability they will amply repay the expense of shed room. They ought also to be kept in the best repair, which may be done at leisure times, so as always to be ready for use. [Some additional remarks on this subject will be found under the head of plows.]

THE AGRICULTURAL EDUCATION OF THE FARMER.

Though last mentioned, this is the first in importance to the farmer's success. It should commence with the thorough, groundwork attainments everywhere to be acquired in our primary schools; and it should embrace an elementary knowledge of mechanics, botany, entomology, chemistry and geology, nor can it be complete without some acquaintance with anatomy and physiology. The learner ought then to have a complete, practical understanding of the manual operations of the farm; the best manner of planting, cultivating and securing crops; he should be familiar with the proper management, feeding and breeding of animals; the treatment of soils, the application of manures, and all the best practices and most approved principles connected with agriculture. This will be but the commencement of the farmer's education, and it should be steadily pursued through the remainder of his life.

He must also learn from his own experience, which is the most certain and complete knowledge he can obtain, as he is thus aware of all the circumstances which have led to certain results; and he should also learn from the experience of his neighbors, and from his personal observation on every subject that comes within his notice. He will be particularly assisted by the agricultural journals of the present day, which embrace the latest experience of some of our best farmers, throughout remote sections of country, on almost every subject pertaining to his occupation. To these should be added, the selection of standard, reliable works on the various topics of farming, and of the latest authorities, which can be procured for direction and reference.

Agricultural colleges and schools should be added to this list of aids to farming, where experienced and gifted minds could be placed, surrounded by every means for conveying instruction in the fullest, yet most simple and effective manner, and with every requisite for practical illustration. It can hardly admit of a doubt, that this neglected field will soon be efficiently occupied, and thus supply the only link wanting to the thorough education of the farmer.

CHAPTER XIII.

~~~~~~~~ ~~~

FARM BUILDINGS.

GREAT neglect is manifest in this country, in the erection of suitable farm buildings. The deficiency extends not only to their number, which is often inadequate to the wants of the farm, but more frequently to their location, arrangement and manner of construction. The annual losses which occur in consequence of this neglect would, in a few years, furnish every farm in the Union with barns and out-houses fully adequate to the necessary demands for both. I will give briefly in detail, the leading considerations which should govern the farmer in their construction.

THE FARM HOUSE.

If this is required for the occupation of the owner, it may

Fig. 84.

be of any form and size his means and taste dictate. If for a tenant, and to be employed solely with a reference to its value to the farm, it should be neat, comfortable and of cor-

Fig. 85.

venient size. It should especially contain a cool, airy and

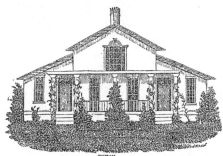

Fig. 86.

spacious dairy room, entirely free from access by any foul
air from any direction; unless the owner prefers one independent of the house, over a clear
spring or cool rivulet, where, partially protected from the sun by a
sheltering bank, half buried in the
earth, and made, as it should be, of
stone, the cool atmosphere within
will afford the best safeguard against
flies and other insects, and preserve
the butter and cheese in the finest
condition.

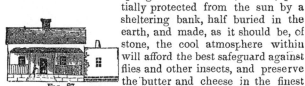

Fig. 87.

Stone or brick are the best materials for dwellings, as
they are cooler in summer and warmer in winter; and if
comfort be the object of the farmer's toil, there is certainly
no place where it should be sooner consulted or more amply
provided for, than in his own home. A naked, scorching

exposure, equally with a bleak and dreary one is to be avoided. The design of a house is protection to its inmates, and if there be no adequate shelter from the elements, it fails in its purpose. It should be tastefully built, as this need not materially increase the expense, while it adds a pleasant feature to the farm. It ought to occupy a position easily

Fig. 88.

accessible to the other buildings and the fields, and yet be within convenient distance of the highway. It is desirable to have it so far removed as to admit of a light screen of trees, and nature will thus add an ornament and protection in the surrounding foliage, which no skill of the architect can equal.

Fig. 89.

Note. For cuts 84, 85, 88, and 91, the author is indebted to A. J. Downing, Esq.

Fig. 84 is a plain house, occupying an elevated position, with a slight drapery of trees and shrubbery attached.

Fig. 85 is the same house and grounds as the former, with the house altered to conform to the Orné or Gothic style, which has recently come into vogue.

Figs. 86 and 87 are plain but neat and pretty cottages,

FIG. 90.

many of which, nearly similar in appearance, are seen throughout the country.

Figs. 88 and 89, of the Orné style, are of about equal pre-

FIG. 91.

tensions and cost.

Fig. 90 is a more imposing country cottage, in the Grecian style; and fig. 91, a pretty lodge, or tasteful cottage for a small family.

THE CELLAR.

This is an essential appendage to a house, particularly where roots are to be stored. Many appropriate a part of it to the dairy, and if thus employed, it should be high, clean and well ventilated, and wholly free from all earthy smell or odor of any kind. The proper preservation of what is contained in it, and the health of the inmates, demand a suitable dryness and free circulation of air. The cellar is frequently placed on the side of a hill, which renders it more accessible from without. This is in no respect objectionable, if the walls are made sufficiently tight to exclude the frosts. When on level ground, it should be sunk only three or four feet below the natural surface, and the walls raised high enough to give all the room wanted; and the excavated earth can be banked around the house, thus rendering it more elevated and pleasant. It also provides for the admission of light and air through small windows, which are placed above ground. A wire gauze to exclude flies, ought to occupy the place of the glass in warm weather, and if liable to frosts, there should be double sashes in winter.

Ventilation is important at all times, and it may be secured even in winter, by a large aperture connected with the chimney. This may be increased in mild weather or during the warmer part of the day, by throwing open the windows. The cellar should be connected with the kitchen or sheds above, by safe, well-lighted stairs; and the entire building should be rat-proof. This is easily accomplished. When erecting a building, the carpenter and mason, for less than the additional expense of a year's support for a troop of rats, can forever exclude them from it, by the exercise of a little ingenuity and trouble. A brick floor in a cellar is easily broken up by these insidious, ever-busy vermin; and a plank or wooden floor is objectionable, from its speedy decay. The most effective and permanent barrier to their inroads, is afforded by a stone pavement, laid with large pieces in cement, closely fitted to each other and to the side walls. This is also secured by placing a bed of small stones and pebbles on the ground and *grouting*, or pouring over it a mortar made of lime and sand, so thin as to run freely between the stones. When dry, a slight coating of water-lime cement is added, which is smoothed over with the trowel. This can be so laid as to admit of ready and perfect drainage, by a depression in the centre or sides, which answers for gutters.

THE BARN

Is the most important addition to the farm. Its size and form and manner of construction must depend on the situation, the means of the owner, and the purposes for which it is designed. It is sometimes essential to have more than one on the premises, but in either case, they should be within convenient distance of the house. They ought to be large enough to hold all the fodder and animals on the farm. Not a hoof about the premises, should be required to brave our northern winters, unsheltered by a tight roof and a dry bed. They will thrive so much faster and consume so much less food when thus protected, that the owner will be tenfold remunerated for the expense necessary to accomplish this object. Disease is thus often prevented, and if it occurs, is more easily removed. The saving in fodder by placing it at once under cover when cured, is another great item of consideration. Besides the expense of stacking and fencing, the waste of the exposed hay in small stacks, is frequently one fourth of the whole, and if carelessly done, it will be much greater. There is the further expense of again moving it to the barn, or foddering it in the field which greatly increases the waste.

It is a convenient mode, to place a barn on a side hill inclining to the south-east, whenever the position of the ground admits of it. There are several advantages connected with this plan. Room is obtained by excavation and underpinning, more cheaply than in the building. An extensive range of stabling may be made below, which will be warmer than that afforded by a wooden building, and the mangers are easily supplied with the fodder stored above. Cellar room can be made next to the bank, in which all the roots required for the cattle can be safely stored, in front of their mangers, and where they are easily deposited from carts, through windows arranged on the upper side, or scuttles in the barn-floor above. More room is afforded for hay, in consequence of placing some of the stables below, and in this way a large part of the labor of pitching it upon elevated scaffolds is avoided. The barn and sheds ought to be well raised on good underpinnings, to prevent the rotting of sills, and to allow the free escape of moisture, as low, damp premises are injurious to the health of animals.

Fig. 92 is a barn placed on a side hill, which is a type of many we see throughout the northern States. The under ground space may be used either as a stercorary (a place for housing manures), or as stables for the cattle or sheep

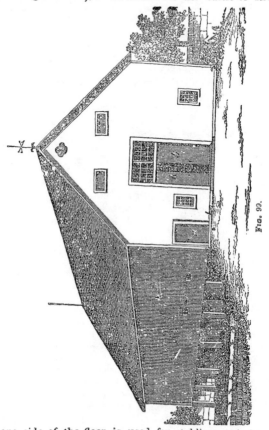

Fig. 99.

But one side of the floor is used for stabling, and the remainder is devoted to the storage of hay, grain, roots, or other cattle food; or it may be employed for storing wagons plows, and other farm implements. Barns of this style are among the most convenient of the farm buildings.

Fig. 100 is an end, and Fig. 101, a side view of an immense barn, capable of holding 100 tons of hay, and 100 head of cattle and young stock.

Fig 102 is a ground plan of the same building, with **two** .

FIG. 100.

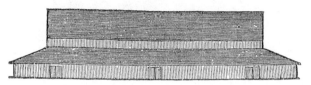

FIG. 101.

sheds attached at each end; *a*, main floor; *b*, *b*, mows for storing hay; *c*, *c*, *c*, *c*, *c*, *c*, stables; *d*, *d*, *d*, *d*, passage ways;

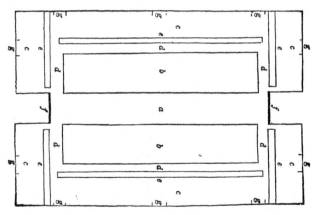

FIG. 102.

e, *e*, *e*, *e*, *e*, *e*, mangers for feeding; *g*, *g*, *g*, *g*, *g*, doors. This gives a large amount of room for the animals, forage, grain, or implements, in a small compass.

Fig. 103 is an enclosed shed, suitable for cattle when the weather is not too cold, and if boarded more closely, would at all times afford desirable accommodations for sheep. It

is not unlike many southern barns; though these are more
frequently made much wider, with the centre enclosed for

Fig. 103.

holding the grain and fodder. The roofs projecting far out
on either side, furnish shelter for the mules, horses and oxen.

Saving all the manure is one of the most important
considerations in the arrangement of the barns, sheds and
yards. The stables should have drains, that will carry off
the liquid evacuations to a muck-heap or tank, and whatever
manure is thrown out, should be carefully protected. The
manure contains the future crops of the farmer, and unless
he is willing to forego the latter, he must carefully husband
the former. A low roof, projecting several feet over the manure which is thrown from the stables, will do much to
prevent waste from sun and rains. The eaves must be
supplied with tight troughs to carry off the water, which
may all be saved for the use of the stock, by leading it into
tight cisterns or reservoirs.

The mangers ought to be so constructed as to econo-
mize the fodder. Box-feeding for cattle, I prefer, as in
addition to hay, roots and meal may be fed in them without
loss; and with over-ripe hay, a great deal of seed may in
this way be saved, which will diminish the quantity necessary to be purchased for sowing. The fine leaves and small
fragments of hay are thus kept from waste, which in racks,
are generally lost by falling on the floor. Racks are objectionable, unless provided with a shallow box underneath,
and they are especially to be avoided in foddering in the
open yards. There is a loss in dragging the forage to them ;
and too often this is done near a herd of hungry cattle, that
gore each other, and are scarcely to be kept at bay by the

use of the stoutest goad. There is also a waste of the hay that falls while the cattle are feeding, and this is largely increased in muddy yards; added to which, the animals are exposed to whatever bad weather there may be while eating, which is at all times to be deprecated.

SHEDS.

Feeding in sheds is far better, and in many instances may take the place of the stall or stable. They are frequently and very properly arranged on two sides of the cattle yard, the barn forming a third, and the fourth opening to the south, unless this is exposed to the prevailing winds. This ar rangement forms a good protection for the cattle ; and connecting the sheds with the barn is of importance in economizing the labor in foddering. The racks or boxes are placed on the boarded side of the shed, which forms the outer side of the yard, and they are filled from the floor overhead. If the space above is not sufficient to contain the necessary quantity of fodder, it may be taken from the mows or scaffolds of the barn, and carried or dragged over the floor to the place wanted. The floors ought to be perfectly tight to avoid waste, and the sifting of the particles of hay or seed on the cattle and sheep.

Unless the ground under the shed be quite dry, it is better to plank it, and it will then admit of cleaning with the same facility as the stables. A portion of the shed may be partitioned off for close or open stalls, for colts, calves or infirm cattle, and cows or ewes that are heavy with young. A little attention of this kind, will frequently save the life of an animal, and add much to their comfort and the general economy of farm management. The surplus straw, corn-stalks, and coarse hay can be used for bedding, though it is generally preferable to have them cut and fed to the cattle.

WATER FOR THE CATTLE YARD

Is an important item, and if the expense of driving the animals to a remote watering place, the waste of manure thereby occasioned, the straying of cattle, and sometimes loss of limbs or other injury resulting from their being forced down icy slopes or through excessive mud, to slake their thirst—if all these considerations are taken into account, they will be found annually to go far towards the expense of supplying water in the yard, where it would at all times be accessible All animals require water in winter, except

such as have a full supply of roots ; and though they some‧
times omit going to distant and inconvenient places where
it is to be had, they may, nevertheless, suffer materially for
the want of it.   When it is not possible to bring water
into the yard from a spring, or by means of a *water ram*,
or it is not easily reached by digging, an effectual way of
procuring a supply through most of the year is by the construction of

Cisterns.—Where there is a compact clay, no further preparation is necessary for stock purposes, than to excavate the
cistern of a sufficient size and depth ; and to keep up the
banks on every side, place two frames of single joist around
it, near the top and bottom, between which and the banks,
heavy boards or plank may be set in an upright position.
The earth keeps them in place on one side, and the joist
prevents their falling.   They require to be only tight enough
to keep the clay from washing in, as no appreciable quantity of water will escape from the sides or bottom.   I have
used such for years, without repairs or any material waste
of water.   They should be made near the buildings ; and
the rains carefully conducted to them by the eaves-troughs
and pipes from an extensive range, will afford an ample
supply.

For household purposes, one should be made with more
care and expense, and so
constructed as to afford
pure filtered water at all
times.   These may be
formed in various ways,
and of different materials,
stone, brick, or even
wood; though the two
former are preferable.
They should be perma-
nently divided into two
apartments, one to re-
ceive the water, and an-
other for a reservoir to

Fig. 104.

contain such as is ready for use.   Alternate layers of gravel,
sand, and charcoal at the bottom of the first, and sand and
gravel in the last, are sufficient; the water being allowed to
pass through the several layers mentioned, will be rendered
perfectly free from all impurities.   Some who are particularly choice in preparing water, make use of filtering stones,
but this is not essential.   Occasional cleaning may be ne-

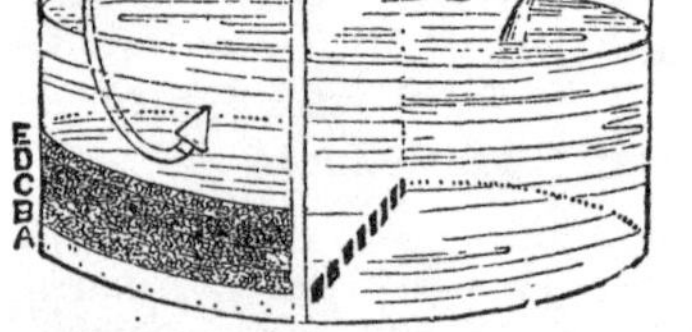

cessary, and the substitution of new filtering materials will at all times keep them sweet.

THE CARRIAGE HOUSE, STABLE AND GRANARY.

The carriage house and horse stable sometimes occupy a building distinct from the barns and other outhouses, which is a good precaution against fire; and where this is the case, it is frequently convenient to have the upper loft for a granary. The propriety of having this proof against rats is obvious. Yet it should be capable of thorough ventilation, when the grain is damp or exposed to injury from want of air. Entire cleanliness of the premises is the best remedy against weevil and other noxious insects.

The *corn-crib.*—If there be more Indian corn on the premises than can be thinly spread over an elevated, dry floor, it may be stored in a corn-crib. This ought to occupy an isolated position; and must be made with upright lattice-work, and a far projecting roof, with the sides inclining from a vertical line towards each other, from the roof downwards, to avoid the admission of rain. The corn in the cob is stored in open bins on either side, leaving ample room in the centre for threshing, or the use of the corn-sheller. Close bins may occupy the ends for the reception of the shelled grain. All approach from rats and other vermin may be avoided, by placing the building on posts, with projecting stones or sheet iron on the top, and so high that they cannot reach it by jumping.

A TOOL HOUSE AND WORK SHOP

Ought always to have a place about the premises. In this building, all the minor tools may be arranged on shelves, or in appropriate niches, where they can at once be found, and are not exposed to injury or theft. Here too the various farming tools may be repaired, which can be done in those leisure intervals that often occur.

Ample shed room for every vehicle and implement about the farm should always be provided. Their preservation will fully repay the cost of such slight structures, as may be required to house them. A wagon, plow, or any wooden implement, will wear out sooner by exposure to all weathers, without use, than by careful usage with proper protection.

A horse power, either stationary or movable, can be made to contribute greatly to the economy of farming operations, where there is much grain to thresh, or straw, hay or corn-stalks to cut. With the aid of this, some of the portable mills may crush and grind much of the grain required for feeding. Even the water may be pumped by it into large troughs for the use of cattle, the fuel sawed, and various other operations performed, which may add much to the convenience of the farmer and save more expensive labor.

A STEAMING APPARATUS.

Where there are many swine to fatten, or grain is to be fed to cattle or horses, this is at all times an economical appendage to the farm. It has been shown from several experiments, that cattle and sheep will generally thrive as well on raw as on cooked roots ; but horses do better on the latter, and swine will not fatten on any other. For all animals excepting store sheep, and perhaps even they may be excepted, grain or meal is better when cooked. Food must be broken up before the various animal organs can appropriate it to nutrition ; and whatever is done towards effecting this object before it is fed to the stock, diminishes the necessity for the expenditure of vital force in accomplishing it, and thereby enables the animal to thrive more rapidly and do more labor, on a given amount. For this reason, I apprehend, there may have been some errors undetected in the experiments in feeding sheep and cattle with raw and cooked roots, which result in placing them apparently on a par as to their value for this purpose.

The crushing or grinding of the grain insures more perfect mastication, and is performed by machinery at much less expense, than by the animals consuming it. The steaming or boiling is the final step towards its easy and profitable assimilation in the animal economy. With a capacious steaming-box for the reception of the food, the roots and meal, and even cut-hay, straw and stalks may be thrown in together, and all will thus be most effectually prepared for nourishment. There is another advantage derivable from this practice. The food may at all times be given at the temperature of the animal system, (about 98° of Farenheit). and the animal heat expended in warming the cold and sometimes frozen food, would be avoided.

The steaming apparatus is variously constructed. I have used one consisting of a circular boiler five and a half feet long by twenty inches diameter, made of boiler iron and laid lengthwise on a brick arch. The fire is placed underneath and passes through the whole length and over one end, then returns in contact with the boiler, through side flues or pockets, where it enters the chimney. This gives an exposure to the flame and heated air of about 10 feet. The upper part is coated with brick and mortar to retain the heat, and three small test cocks are applied at the bottom, middle and upper edge of the exposed end, to show the quantity of water in it; and two large stop cocks on the upper side for receiving the water and delivering the steam, completes the boiler.

The steaming-box is oblong, seven or eight feet in length, by about four feet in depth and width, capable of holding 60 or 70 bushels, made of plank grooved together, and clamped and keyed with four sets of oak joist. A large circular tub, strongly bound by wagon tire and keyed, and holding about 25 bushels is also used. The tops of both are securely fastened, but a two inch auger hole, protected by a leather valve, permits the escape of any excess of steam. The steam is conveyed from the boiler into these, by a copper tube attached to the steam delivery-cock, and it is continued into the bottom of the box and tub by a lead pipe, on account of its flexibility, and to avoid injury to the food from the corrosion of the copper. It is necessary to have the end of the pipe in the steaming-box, properly guarded by a metal strainer, to prevent its clogging from the contents of the box.

FIG. 103.

I find no difficulty in cooking 15 bushels of unground

Indian corn in the tub, in the course of three or four hours, and with small expense of fuel. Fifty bushels of roots can be perfectly cooked in the box, in the same time. For swine, fattening cattle and sheep, milch cows and working horses, and perhaps oxen, a large amount of food may be saved by the use of such or a similar cooking apparatus. The box may be enlarged to treble the capacity of the foregoing, without prejudicing the operation, and even with a boiler of the same dimensions, but it would take a longer time to effect the object. If the boiler were increased in proportion to the box, the cooking process would of course be accomplished in the same time.

Fig. 105 is a good form of a steaming apparatus, essentially similar in principle to the one described.

ICE HOUSES.

These, in the rapid progress of improvement and the increasing comforts and luxuries of this country, are justly deemed an important addition to the farm buildings. They are frequently essential to the operations of the dairy, and the preservation of milk, butter and cream for a longer time, and in better condition, than is otherwise attainable. They are also useful for keeping meats, fruits, eggs and vegetables unchanged, for an almost indefinite period. Whether this is done with reference to sending the articles to market at the most convenient or advantageous time, or to their consumption at home, it is equally consistent with economy.

It is not necessary to dig into the earth for the purpose of securing a good ice house. Indeed, a large quantity of ice can be stored more cheaply by constructing the ice house above the surface of the ground. The main object is to secure *isolation of the ice, and surround it with an adequate barrier of non-conducting materials.* To do this effectually, a triple wall of plank or boards must be made, from six to eighteen inches apart, and the spaces between each compactly filled with straw or tan bark. The bottom must be equally well secured, and have drains for the escape of the water, yet not for the admission of air. The top has a double roof, and a thick coating of straw is spread over the ice.

The preservation of ice depends, in addition to the foregoing, on the observance of principles, of which many are entirely neglectful. There should be no access to the ice except on the top, and the sides and ends must be perfectly tight. Cold air being heavier than warm, in the ratio of its diminished temperature; it follows, that air which is near the freezing point, if sustained by tight walls, cannot be displaced by warmer air, unless a current is forced upon it from without, which must be avoided. As well may mercury be driven out of a vessel by pouring water into it. The cold air settles upon the ice and remains there permanently, and of course the ice has no tendency to melt even during the hottest weather. By adhering to these principles, ice may be kept for years, and almost without waste. It is important to put up ice in the largest, most compact blocks, and in the coldest weather. By the use of large pieces closely wedged together, there is security against the circulation of air through the mass; and by doing this in the severest weather, the ice goes into store with a greatly augmented intensity of cold. Some do not consider the difference in the temperature either of snow or ice, when each are equally susceptible of degrees of temperature below 32°, as the atmosphere, metals, or other substances. If put into the store room at zero, ice must be elevated 32° before it rises to the melting point, and it has, therefore, all this stock of cold (privation of heat) which it must first exhaust, before it assumes the form of a liquid. A rigid observance of the above rules, will preserve ice anywhere, either above or below ground.

The materials for farm buildings I have assumed to be of wood, from the abundance and cheapness of this material generally in the United States. Yet when not too expensive, or where capital can be spared for this purpose, brick or stone should always take their place. They are more durable, less exposed to fire, and they sustain a more equable temperature in the extremes of the seasons.

Barns and sheds cannot, like houses, be conveniently made rat proof, but they may be so constructed as to afford them few hiding places, where they will be out of the reach of cats and terriers, which are indispensable around infested premises. These and an occasional dose of arsenic, carefully and variously disguised, will keep their numbers within moderate bounds. If poison be given, it would be well to shut up the cats and terriers for three or four days until the object is effected, or they, too, might partake of it.

LIGHTNING RODS.

During the sultry weather of American summers, thunder showers are frequent and often destructive to buildings. This danger is much increased for such barns as have just received their annual stores of newly cut hay and grain. The humid gases, generated by the heating and sweating of the hay, which immediately follows its accumulation in closely-packed masses, offers a strong attraction to electricity, just at the time when it is most abundant. It is an object of peculiar importance to the farmer, to guard his buildings, at such times, with properly constructed lightning rods ; and they are a cheap mode of insurance against fire from this cause, as the expense is trifling and the security great.

It is a principle of general application, that a rod will protect an object at twice the distance of its height above any given point, in a line perpendicular to its upper termination. Thus a rod attached to one side of a chimney of four feet diameter, must have its upper point two feet above the chimney to protect it. Its height above the ridge of a building, must be at least one half the greatest horizontal distance of the ridge from the perpendicular rod.

Materials and manner of construction.—The rod may be constructed of soft, round or square iron (the latter being preferable), in pieces of convenient length, and not less than 3-4ths of an inch in diameter. These should not be hooked into each other, but attached either by screwing the ends together, or forming a point and socket to be fastened by a rivet, so that the rod when complete, will appear as one continuous surface of equal size throughout. If a square rod be used, it will attract the electricity through its entire length, if the corners be notched with a single downward stroke of a sharp *cold chisel*, at intervals of two or three inches. Each of these will thus become a point to attract and conduct the electricity to the earth. A bundle of wires, thick ribbons, or tubes of metal, are more efficient conductors, than an equal quantity of matter in the solid round or square rods, as *the conducting power of bodies, is in the ratio of their surfaces.* No part of the rod must be painted, as its efficiency is thereby greatly impaired. The upper extremity may consist of one finely-drawn point, which should be of copper or silver, or well-gilded iron, to prevent rusting. The lower part of the rod, at the surface of the ground, should terminate in two or three flattened, divergent branches, leading several feet outwardly from the building, and buried at the depth of perpetual moisture, in a bed of charcoal. Both the charcoal and moisture are good conductors, and will ensure the passage of the electricity

Into the ground, and away from the premises. The rod may be fastened to the building by glass or well-seasoned wood, boiled in linseed oil, then well baked and covered with several coats of copal varnish.

The conductors of electricity in the order of their conducting power, are copper, silver, gold, iron, tin, lead, zinc, platina, charcoal, black lead (plumbago), strong acids, soot and lampblack, metallic ores, metallic oxides, diluted acids saline solutions, animal fluids, sea water, fresh water, ice above 0°, living vegetables, living animals, flame, smoke, vapor and humid gases, salts, rarified air, dry earth, and massive minerals. The non-conductors in their order, are shellac, amber, resins, sulphur, wax, asphaltum, glass, and all vitrified bodies, including crystalized transparent minerals, raw silk, bleached silk, dyed silk, wool, hair and feathers, dry gases, dry paper, parchment and leather, baked wood and dried vegetables.

Palladium, reckoned among the noble metals, was first discovered by Dr. Wollaston, in 1803. It has been found to possess a conducting power about 50 per cent. greater than copper, and consequently, is the best conductor known. This metal is not liable to oxidation, and is therefore, in every respect the most desirable material for the points or upper extremity of the rods, and for this purpose it has recently been extensively introduced.

It will be seen above, that water is a tolerable conductor of electricity, and when exposed to rain, all the non-conductors are liable to become temporary conductors. To secure them as non-conductors, it is therefore necessary to protect them from contact with rain or moisture.

Rust or the oxides of metals, destroy their conducting power, and to secure them as conductors, it is essential to keep the rods free from rust, paint, oil or varnish, leaving nothing on the exterior but the pure metal.

INDEX.